Carlos Emilio Padilla Severo

Technology and Mediation in Distance Education

Carlos Emilio Padilla Severo

Technology and Mediation in Distance Education

Expanding educational possibilities in online environments with technology to support pedagogical mediation

ScienciaScripts

Imprint

Any brand names and product names mentioned in this book are subject to trademark, brand or patent protection and are trademarks or registered trademarks of their respective holders. The use of brand names, product names, common names, trade names, product descriptions etc. even without a particular marking in this work is in no way to be construed to mean that such names may be regarded as unrestricted in respect of trademark and brand protection legislation and could thus be used by anyone.

Cover image: www.ingimage.com

This book is a translation from the original published under ISBN 978-613-9-65104-7.

Publisher:
Sciencia Scripts
is a trademark of
Dodo Books Indian Ocean Ltd. and OmniScriptum S.R.L publishing group

120 High Road, East Finchley, London, N2 9ED, United Kingdom
Str. Armeneasca 28/1, office 1, Chisinau MD-2012, Republic of Moldova, Europe
Printed at: see last page
ISBN: 978-620-7-84913-0

DEDICATORY

I dedicate this thesis to my beloved wife Elenara and our beloved sons Otávio and Gustavo.

I would also like to thank my parents Emílio and Marlene for their constant support and affection.

Index

CHAPTER 1

INTRODUCTION

Pedagogical mediation is an important element in the process of building knowledge. In it, the teacher becomes an essential agent during the development of learning, since he or she must intervene with the aim of promoting learning and student reflection. This reflection aims to promote the exchange of ideas and awaken curiosity in order to find solutions to the proposed didactic-pedagogical activities.

As a mediator of learning, the teacher not only plays the role of facilitator in the process of constructing knowledge by implementing strategies aimed at circumventing problems in the acquisition of new concepts, but must also act to help the student acquire autonomy (WERTSCH, 1999). Autonomy is the basis for students to develop a sense of analysing and organising content, as well as developing the ability to select the most significant content related to the knowledge they want to acquire.

Student autonomy aims to promote the construction of learning that allows the individual to self-regulate (DIAS et. al., 1993), which is based on associating new concepts with previously known content. Pedagogical mediation is a central concept within an approach that aims to break away from education based on teaching or the mere transmission of content. It is an approach that seeks to conceive of a more participative, creative individual who is capable of relating concepts to form more solid knowledge based on previously assimilated information. It is a student-centred approach, aimed at student autonomy through motivation and the acceleration of maturity (GUTIERREZ and PRIETO, 1991).

During mediation, the teacher will develop strategies to overcome possible problems and adapt to the different needs and particularities of each student, in pursuit of their pedagogical intentions. This premise is valid in both face-to-face and distance learning. However, the particularities and differences inherent in face-to-face and virtual teaching spaces mean that mediation is not a mere transposition. As such, this research aims to address issues related to the support of mediation in online education environments using computer technologies.

CHAPTER 2

DELIMITATION OF THE OBJECT OF STUDY

This chapter presents the elements that make up the context of this study, listing the conceptual aspects that make it possible to justify and characterise the research covered in this thesis. The focus of this thesis was to find evidence of pedagogical mediation in a virtual teaching-learning environment. In this research, we monitored the mediation that took place in the *online* educational environment, based on the[1] textual interactions carried out by the participants in the environment's textual interaction tools. With this monitoring, we aimed to identify the level of mediation of student learning. To do this, we used text mining technology to help map the categories of interactions carried out by the participants. With this, we can identify the levels of mediation based on the participants' interactions. Therefore, in the next few paragraphs we will discuss aspects that allow the research to be contextualised.

Distance education is a type of teaching that allows spaces to be created for the generation, promotion and development of situations in which students, teachers and tutors learn; and the mediation of relationships between teachers, tutors and students is a distinctive feature (LITWIN, 2001). According to Moran (2003), in *online* education the teacher is faced with a multiplicity of distinct and complementary roles, where the teacher's need for adaptation and creativity increases in the face of new situations, pedagogical proposals and activities. In addition, distance education, as a teaching modality, has always been defined by the substantive incorporation of different media, since teaching does not take place in conventional classroom spaces, which gives students a variety of possibilities for learning (LITWIN, 2005).

In general, monitoring student participation in activities in virtual teaching-learning environments (VLEAs) provides the teacher with information

---

[1] Moore and Kearsley (2007) define interaction from three perspectives: student-content, student-tutor and student-student. According to the authors, interaction occurs through different genres of communication. The focus of this research was on interactions between student-tutor and student-student.

for the pedagogical intervention needed to develop learning. This stage of the teaching-learning process, similar in functionality to its face-to-face counterpart, is fundamental in distance education, providing elements that help the teacher in assessment by monitoring the evolution of each student's learning process. However, in the way that an AVEA is currently structured, collecting this information is a major challenge. This challenge arises from the voluminous information generated from user interactions with the most diverse tools available.

Furthermore, the monitoring of student participation is based more on quantitative than qualitative aspects. This can be seen in the research carried out on the subject. As in the work by Ferreira et. al. (2003), which presents an interface for recording and analysing assessments made by the teacher during the development of a course. Romani (2000) presents a tool to help monitor interactions in AVEAs, by mapping students and students with teachers, enabling the quantification and graphical visualisation of relationships between individuals based on their interactions. Bassani and Behar (2006) focus on quantifying accesses and the number of contributions posted in an AVEA, presenting reports on each student's interactions.

> Bearing in mind the potential of the tools in a virtual environment as instruments[2] of interaction, as well as their particularities and influences on the development of student learning, pedagogical mediation is an important activity that emphasises the role of the teacher and tutor as agents of integration and guidance of technological resources in improving the use of a virtual environment as an instrument to support teaching and learning.

The effective participation of students in teaching activities is essential for the development of learning, something that can be observed in traditional face-to-face teaching. However, in a teaching activity carried out in a virtual environment, verification of student participation is limited to records of their interactions with the environment, or with other individuals participating in an

---

[2] An instrument is understood here as a means by which man transforms nature (VYGOTSKY, 2007).

interaction session, both synchronously and asynchronously. Virtual teaching-learning environments store subjects' interactions with the tools in an eminently textual form, such as *chats,* forums, *emails,* portfolios and diaries. A lot of information can be obtained from students' interactions with the virtual environment, but we are only interested in information that is relevant to the pedagogical mediation process. Therefore, some mechanism for mapping these interactions would be appropriate and desirable, so that the teacher can use information and communication technologies as a tool to aid the learning mediation process.

Given that observing and analysing the content of student interactions in the virtual environment can be very stressful, depending on the number of students taking part in a course (GLUZ et. al., 2008). The research focus of this study centred on ways of mitigating work overload during pedagogical mediation in virtual teaching-learning environments.

In this thesis, we investigated how to build and apply a mining system that helps reduce the teacher's workload during their pedagogical mediation activities in *online* educational environments. This process sought to provide support to help the teacher identify the level of mediation of student learning, so that they can identify those students with the greatest need for pedagogical intervention during the development of an *online course.*

In order to understand mediation as a pedagogical process, we used socio-historical theory as the epistemic basis for this research, through the works of Vygotsky (2007; 1998), as well as contemporary authors along the same lines such as Werstch (1998), Cole (1990), Kozulin (1998), Gallimore and Tharp (1996), Moll (1993) and Diaz et. al. (1993), who represent the foundation for studies on mediation activities in VLEA. The choice of epistemological approach is related to the affinities of the research group, as well as the theoretical principles that drive the investigations of the MEDIATEC research project, in which this doctoral thesis is inserted. MEDIATEC is a project financially supported by CNPQ that aims to see how pedagogical mediation adjustments can be made in the educational context of distance learning. The aim is to investigate how teacher mediation in virtual teaching and learning environments can be shared by a technology that supports pedagogical intervention activities, with the aim of easing the teacher's workload. To this end, the MEDIATEC project involves interdisciplinary work that combines elements from the fields of social interaction, distance education, intelligent systems, machine learning and cognitive models. The research has already produced some significant results, which are described in the works of (GLUZ et. al., 2008) (PASSERINO et. al., 2007a; 2007b) (KOCH at. al., 2009) (RAMINELLI et. al., 2009) (SEVERO et. al., 2009a; 2009b).

Therefore, the concept of mediator proposed by Kozulin (1998) is used here, in which the mediator is seen as the person who interacts with the individual learner by monitoring the development of their cognitive functions, seeking to organise thought and improve learning processes. According to Werstch (1998), mediation occupies a central theoretical role in Vygotsky's work, especially semiotic mediation. According to socio-historical research, human beings have access to the world in an indirect or mediated way, both in the way we obtain information and in the way we act on it through our actions. According to Daniels (2003), the mediator plays a fundamental role, as it is the means by which the individual acts on social, cultural and historical factors and suffers their action.

Therefore, in this thesis we use the term mediator as the individual who will intervene pedagogically through the use of mediation instruments in the form of tools in a virtual teaching and learning environment. Thus, pedagogical mediation will be the action developed by the teacher in interaction with their students within a space of interaction that we will call a virtual teaching-learning environment.

Based on the identification of mediation categories[3] , we try to find out what level of the mediation process the students are at, so that we can help the teacher identify the degree of student autonomy in the development of their learning. The work of Diaz (1993) and later Passerino (2005) provided the

---

[3] It should be noted that these categories of mediation will be presented in more detail in a specific section of this thesis.

conceptual support for identifying these categories of mediation. In a nutshell, the mediation categories are ways of verifying the degree of intervention required by the teacher during their teaching activities and the degree of student knowledge based on their interactions in the environment.

Preliminary research has shown that few studies relate socio-historical theory to the area of computing in terms of finding evidence of mediation in virtual environments. The work by Andrade and Vicari (2003) explored the theme of the Zone of Proximal Development in the conception of a student model in virtual teaching-learning environments, associating Artificial Intelligence techniques such as pedagogical agents and mentalist models. Faria's work (2003) analysed interactivity in virtual teaching-learning environments, whose investigation sought to uncover how relationships, communications and mediations were established between individuals in the environment; however, it did not establish any relationship between the use of computer technology and the mediation process. Machado's work (2009) investigated the relevance of pedagogical mediation in virtual teaching-learning environments from a socio-historical perspective, but also did not associate information and communication technology with pedagogical mediation. Silva (2008) uses the principles of interaction and pedagogical mediation from a socio-historical perspective to investigate the influence of hypertextual genres on interaction and pedagogical mediation between individuals involved in distance learning courses over the Internet.

In this research, we emphasise the importance of combining epistemological principles of pedagogical mediation in virtual environments with computer technologies that help the teacher's didactic-pedagogical activities by surveying evidence of mediation in such environments. To this end, it was necessary to use text mining technology to build a support system for the mediation process. Many works found in the literature reinforce this idea, such as Xu et. al. (2005) and Xu and Wang (2006) who explored the use of data mining as an appropriate technique for designing personalised virtual learning environments. Penã et. al. (2002) presented a multi-agent environment for personalised assistance in navigation strategies, content selection and navigation tools for students in virtual environments. Schiafino et. al. (2008) explored the use of intelligent agents for personalised assistance to students during their activities in a virtual teaching and learning environment. Casamayor et. al. (2009) presented a proposal to use intelligent agents to resolve conflicts resulting from communication and collaboration in virtual environments by monitoring student interactions in the environment to provide personalised assistance to the teacher.

The works cited were focused on analysing and building student profiles, describing learning styles based on their interactions with tools in a virtual teaching and learning environment. These studies did not explore mediation as a social process, in accordance with the epistemological concept addressed in this thesis.

Thus, once the research theme had been identified, questions were drawn up to help identify the problem being addressed and to draw up the assumptions and objectives of the research. Therefore, the following sections will present the factors that justified and motivated the development of this research. As well as the aspects that characterise this work as a doctoral thesis. In addition, the delimitation, strategy and organisation of the study will be presented.

2.1 Problem and research questions

The investigation of the possibilities that Information and Communication Technologies (ICT) offer in terms of supporting mediation in VLEs is the starting point for the thesis presented here. The initial premise is that mining techniques combined with an appropriate epistemological basis can offer unique potential for identifying mediation processes and levels. Thus, some research questions were raised, which served as a guiding basis for identifying and delimiting the research problem:

I. How can the mediation process be perceived and managed in the different interaction spaces, especially forums, *chats* and *wikis, of* a virtual teaching-learning environment?

II. How can text mining be used to map interactions and identify signs of mediation in virtual teaching-learning environments?

III. How can information from mapping be made available to help teachers with the task of pedagogical mediation in a virtual teaching-learning environment?

From the identification of questions that instigate curiosity and motivate investigation and the search for answers, we arrive at the identification of the research problem that we intend to study:

IV. How can text mining technology and socio-historical epistemology provide elements to support pedagogical mediation in virtual teaching-learning environments?

Taking into account the object of study presented and according to the definitions of Oliveira (2000), this research is classified as an applied science approach, since it aims to explore and experiment with knowledge about phenomena of a technological and socio-cultural nature.

In this way, we can identify the interdisciplinary nature of the research through its relationship with two major areas of knowledge: exact sciences and social sciences. The exact sciences include computing, which will support the research through information and communication technologies. The social sciences include education and psychology, whose influence on the thesis lies in the theoretical basis of pedagogical mediation and learning.

In terms of relevance, we believe that the research is unique in that it applies the theoretical aspects involved in learning mediation in an environment that is different from the conventional face-to-face classroom. It is an investigation that seeks to identify and manage the process of pedagogical mediation in the distance learning modality, supported by virtual teaching-learning environments.

Originality, according to Martins and Lintz (2000), refers to the search for new results that have not yet been disseminated in scientific or professional circles. In preliminary research, no mechanism has been identified to support pedagogical mediation in current virtual teaching-learning environments, which provides the mediator with indicative elements about the development of student learning, based on mapping the mediation dynamics resulting from textual interactions between course participants in the *online* environment.

2.2 Delimitations and scope of the study

The delimitation of the research work is essential for any research project, as it defines the extent, depth and type of research approach. These aspects are now relevant, as their definitions prevent the results of the study from being superficial, or even too broad, making the development of the research unfeasible (ECO, 2010). The narrower the field of enquiry, the greater the security provided to the researcher. These limits were defined according to the object of study of this thesis.

This research work includes elements related to the theme previously identified in the object of study, as well as a focus and depth on aspects related to pedagogical mediation in virtual teaching-learning environments. In this way, the research will focus on the following dimensions: (a) deepening and conceptualising the terms involved in pedagogical mediation and virtual teaching-learning environments; (b) aspects related to interaction in the various tools of the virtual environment; (c) text mining technology to support the mapping of interactions and monitoring of pedagogical mediation in the environment.

This thesis will not deal with the problems involved in pedagogical mediation in distance education in all its dimensions, focusing only on the mediation problems found in virtual teaching-learning environments. Nor will different learning theories be discussed or analysed. We have adopted socio-historical theory as the guiding line for the epistemological basis of the research, but we do not intend to compare it with other approaches.

As for the spatial and geographical dimension of the research, it involves a study based on experiments on a local database of courses made available on the AVEA of the Federal University of Rio Grande do Sul - UFRGS, more specifically on courses of the Postgraduate Programme in Informatics in Education - PGIE. The exploratory studies were carried out on data collected from courses held at the institution. As for the scope of the studies, they are limited to the reality outlined by the nature of the resources, structure and human material involved in the courses held in the programme from which the data comes. In this way, the inferences made during this study portray results that are relevant to the institutional context.

2.3 Identifying objectives

After presenting the problem involved in the studies in this thesis, based on the identification of research assumptions based on previously identified questions, we move on to outline the objectives, which will act as guidelines for obtaining answers to the questions raised. According to Jung (2004): *"[...] the objectives should be drawn up on the basis of what is intended to be achieved. The researcher should not specify how he intends to do it, but only what he wants to achieve by carrying out the work"* (p. 220).

The general objective provides a more comprehensive view of the research topic, making it possible to identify intrinsic aspects of the proposal. The specific objectives act as instruments for applying strategies to achieve the general objective of the research.

In addition, the general objective allows the starting point to be identified, following a vision of segmenting the research problem, while the specific objectives are used to define research parameters. This is the strategy for identifying objectives proposed in this thesis.

The general aim of the research is to investigate how a mining system can map the content of interactions in an AVEA, identifying levels of mediation and aiming to identify elements to support pedagogical mediation for the teacher and tutor of an *online* course.

Based on the general objective of the thesis, the following specific objectives served as guidelines for the research:

1) Identify and study text mining techniques for mapping student mediation

based on interaction sessions in AVEA.

2) Proposing a system to support the teacher's pedagogical mediation in virtual teaching-learning environments.

3) Graphically represent the pedagogical mediation actions extracted from interaction sessions between the teacher mediator and the students.

CHAPTER 3

mediation and learning

The purpose of this chapter is to explain aspects related to the mediation of learning in virtual environments, based on epistemological principles grounded in the socio-historical theory initially proposed by Vygotsky and later expanded by followers (LURIA, 1994) (LEONTIEV, 2005) and various scholars (WERSTCH et. al.., 1998; 1999) (BAQUERO, 1998) (DANIELS, 2003) (SMOLKA et. al., 1998; 2000) (COLE, 1990) (KOZULIN, 1998) (GALLIMORE and THARP, 1996) (MOLL, 1993) (DIAZ et. al., 1993).

A previous study is necessary, since these theoretical concepts served as a conceptual foundation for the present research to be developed through this thesis. The study aims to understand the dynamics associated with mediation actions in spaces provided by virtual teaching and learning environments. As such, this chapter seeks to present concepts from socio-historical theory, relating them to pedagogical mediation actions in virtual teaching-learning environments.

In socio-historical theory, human development is based on the constitution of higher psychological processes through social interaction. The great contribution of the socio-historical theory developed by Vygotsky is the view of man as a biological and cultural being. Human learning is a complex process that involves not only rationality, but also the individual's interactions with their social environment. In this way, according to the socio-historical conception, human beings are constituted by their relationships with the external world. According to Vygotsky, *"[...] within a general process of development, two qualitatively different lines of development can be distinguished, differing in their origin: on the one hand, elementary processes, which are of biological origin; on the other, higher psychological functions, of socio-cultural origin [...]"* (2007, p.61).

According to Baquero (1998), the origin of the human psyche lies in the individual's relationship with their sociocultural environment. In other words, the human capacity to recognise and act in the world is the result of the development of their psychological processes, which are influenced by interpersonal factors. Furthermore, according to the author, the development of psychological processes is culturally organised, and the role of school education stands out. The author emphasises that the development of higher mental processes is not the result of the evolution of elementary functions; on the contrary, they are the result of specific situations experienced by the individual in social life with the use of mediating instruments.

Social-historical theory provides a basis for understanding the characteristics involved in the individual's mental and social development process, through an element called mediation, which forms the link between the object of knowledge and learning. Thus, from the perspective of socio-historical development postulated by Vygotsky (2007), all interaction between the subject and a given object of knowledge is socially mediated and does not occur directly. This social mediation is carried out both in symbolic form, through signs internal and external to the individual, and through instruments or the action of another subject.

Tools are resources used by humans to mediate their actions in nature. Within the concept of tools are the physical artefacts created by human beings, which in some way help them in their actions to transform the environment in which they live. Vygotsky and Luria (1996) explain that the instrument is aimed at the human being's control over nature, in other words, it is externally orientated. Thus, instruments play the role of external aids to support the individual's cognitive development, which are used as elements of orientation and memory activation. It's worth noting that instruments are materialised in what is concrete, real, modifying a situation or performing a specific task. A contemporary example is the mobile phone as an instrument for mediating communication between people.

Signs, which are purely abstract and also called psychological instruments, are mental resources used to form higher psychological structures, which reflect the personal appropriation of objects of knowledge present in the world. This personal appropriation is full of meanings that the individual himself establishes in his mental structure. External symbols, on the other hand, are concrete and represent interpositions between the individual and objects in the world, making meaningful representations of shared ideas and concepts that are present in the environment in which the individual is inserted. Baquero (1998) explains that this personal appropriation is guided by a process of internal (intra-subjective) reconstruction of meanings based on external interactions that the individual has with objects of knowledge.

According to Vygotsky (2007), every elementary form of behaviour presupposes a direct reaction to a problem situation faced by the individual. This problem situation is represented by the expression (S R), where (S) refers to an individual and (R) to an object. However, in operations mediated by signs, a second-order element appears in this relationship, establishing a special function of reverse action, i.e. it acts on the individual in the relationship, causing a transformative effect. This second-order element allows the operation to be completed in an indirect way, also known as a mediated way. Thus, the relationship between the individual (S) and the object (R) occurs through a third element (X) that forms a link between them. This third element consists of a mediated action, which is carried out by another individual.

This element of mediation is represented in the human mental structure through an internalised sign. According to Smolka (2000), the internalisation of a sign is the process of an individual's particular signification of external cultural practices, which are reconstructed in the individual's mental apparatus. Baquero (1998) explains that internalisation is a fundamental process for the development of higher mental processes and consists of transformations from an external to an internal activity, i.e. from an interpersonal to an intrapersonal level. Internalisation is often referred to as appropriation. The use of signs constitutes a form of behaviour characteristic of human beings which stands out from their biological development.

This leads to the creation of types of psychological processes that are historically rooted in culture. The intemalisation of cultural forms of behaviour allows for a qualitative leap in the development process of human psychology. According to Vygotsky (2007), the use of signs represents the great distinction between human and animal psychology. Intemalisation is not merely a process of purely transferring an external cultural process to the interior of the individual, i.e. it is not merely copying a model, but rather a transfer with adaptation, adjustment, or the constitution of an individual meaning of its own, which forms the individual's consciousness.

According to Wertsch (1998), the construct of mediation played a central role in Vygotsky's work, especially semiotic mediation[4] , where mediated action forms the basis for formulating socio-historical research.

Leontiev (2005) makes it clear that direct stimulus-response relationships are not sufficient for higher mental development, and that some intermediary term - activity - is needed to formulate an adequate psychological theory to explain the formation of the human mind. Kozulin (1998) complements the idea of mediation by citing that mental processes are mediated and socially significant activities which are produced by the use of external material tools and internal sign systems, leading to a transformation of the individual's behaviour. Daniels (2003) adds that mediation is a concept in which social, cultural and historical forces play an important role in human development. Mediational means are resources through which the individual acts on social, cultural and historical factors, suffering their action and assimilating their particularities.

For Wertsch (1998), mediation is a dynamic process supported by tools and signs, which play an important role in shaping mediating actions. The tools and signs themselves have no power to do anything on their own, so they only have an impact on the individual's psychological structure when they are used. In this way, as mediation is seen as a dynamic process, the introduction of new tools would lead to the transformation of the psychological process, restructuring the

---

[4] Semiotics is a science that studies cultural events, considering them as systems of signs, i.e. systems of signification. Semiotics focuses on the process of representing concepts from nature or human culture (SANTAELLA, 2004).

entire flow of the individual's mental processes. Complementing the characteristics that determine the dynamism of the mediation process, Cole's (1990) studies show that the socio-cultural context is a factor that acts in the selection of cultural tools, influencing and moulding the form of the mediated action.

According to this theory, learning takes place through internalisation, which involves an external activity that is modified until it becomes an activity internal to the individual (KOZULIN, 1998). From this perspective, knowledge is a social production that arises from human activity, which produces it. Through activity, man transforms nature and creates technical, instrumental and semiotic means to carry out his actions.

From the above, we can see that the relationship between the individual and the object of knowledge is not simply one of interaction, but of dialectic, leading to the internal and individual construction of a representation of the world based on social and cultural processes (VYGOTSKY, 1998). An important element in this perspective of the individual's mental development is language, which is not limited to speech, but to all the forms of interaction/communication that human beings use. Language therefore includes all forms of communication, such as gestures, drawings, written records or any sign that represents actions aimed at dealing with and solving problems.

The communication process takes place from a dialogical perspective (BAKHTIN, 2004), in which any element, of any nature (natural, consumer or technological), can become a sign. This is because the element receives a meaning that goes beyond its own particularities, in other words, it also represents another reality, distorting or assimilating it in a particular way.

Vygotsky (2007) and Bakthin (2004) reinforce the importance of using signs as psychological tools in social relationships, especially in communication relationships between people, where dialogues are established. For them, language in all its forms of expression is considered very important in social relations. According to Bakthin (2004), in order to understand something that is presented to us through language, we must first situate and contextualise the object in the social environment. First of all, we need to situate the subjects involved in the communication (sender and receiver), as well as the words enunciated in the social environment, since it is not the sounds of the words that we hear, but what they express, in other words, their meanings and senses.

For Vygotsky (1998), words only acquire meaning when applied to a context, because only contextualisation provides meaning, which is conditioned by the way in which the individual uses it, according to the mind of each person.

Vygotsky's studies highlight the important role played by language in the formation of human consciousness, since language *is* seen as a regulating element of thought, since when it is intemalised it acts in the complex process of human thought formation. Finally, we emphasise that in the cultural appropriation of meanings, language plays a mediating role, since it is developed in the subject through an inter-psychic process, i.e. the acquisition of language is initially social, in which the individual invokes the external environment to gradually create an internal system of signs.

In the next section we will discuss aspects that influence human psychological development, from the perspective of the studies postulated by socio-historical theory.

2.4 Higher psychological functions

A key element in socio-historical theory is the relationship between development and learning which, according to Vygotsky (2007), is realised through semiotic mediation. According to the author, school learning expands the possibilities for human development, since some development processes are only triggered when the individual interacts with others in a given cultural context. In this way, human psychological development occurs through learning in interactional contexts, initially on an inter-psychological level.

Two key aspects in understanding human psychological development are the lines: actual and potential development.

- The actual level of development is all the functions that the individual is able to perform independently, i.e. without the help of someone more experienced.

- The level of potential development includes all the functions that the individual has not yet fully mastered, but is capable of carrying out with the help of someone more experienced, during their social interaction. Therefore, potential functions are those that are in the process of changing until they are assimilated (intemalised) by the individual.

Vygotsky (2007) explains that the relationship between actual and potential development can be described as the difference between cultural development that has already been achieved and that which is yet to be achieved, based on social relations within specific cultural practices. In this way, a difference, or distance, is established between what the individual can achieve on their own and what they can achieve with the help of another, this difference is called the **zone of proximal development (ZDP).** The general law of development for presents the trajectory of human psychological development from a temporal perspective, in which all of an individual's psychological functions initially appear on a social level and then appear on an individual level. Psychological functions initially appear between people (interpsychological) and then within the individual (intrapsychological).

In this way, development has a social origin, i.e. it occurs from the outside to the inside of the individual, through a strong influence of the culture in which the person is inserted in this process. As such, learning becomes a fundamental factor in the subject's development. Vygotsky's work reinforces the idea that learning is a factor that influences human development, where learning is what determines which path the individual's development takes.

Wertsch (1998) emphasises that there is a correspondence between the individual's level of potential development and the functions of the interpsychological level, since both occur during social interactions. Potential development is linked to the help of others during social interaction, as is interpsychological functioning. Likewise, the actual level of development is corresponded to intrapsychological functioning, since both are directed towards the individual's assimilation of social interactions. It is in the ZDP that the teacher, in their mediation activity, will intervene to help the student learn something new. In this sense, pedagogical mediation is present through teaching aimed at the subject's development.

After explaining that the psychological development of human beings occurs from the social to the individual level, Vygotsky (2007) clarifies how this process takes place. The author calls this process of transferring interpsychological functioning to intrapsychological functioning intemalisation. The process of intemalisation is the reconstruction within the individual of an external action. It should be stressed that intemalisation is not the mere transfer of a certain external activity (found in the social environment) to the subject's internal level. Intemalisation is a process in which the internal psychological plane is formed in the individuality of the subject. When mental processes are intemalised, they are transformed to meet the specific needs of each subject's individual system, so they are not mere copies of the functions initially found in the social environment.

> Collaborating with the concept of intemalisation, Smolka (2000) illustrates that it is a form of mastery of cultural ways of acting, thinking and relating socially and with oneself. The actions to be intemalised occur on an intersubjective level based on social relationships and not on the social other, i.e. it is not a simple transfer of mental functioning from individual to individual, but rather a process of appropriation of the higher mental functioning present in the social relationships established in historical-cultural achievements (SMOLKA et. al., 1998).

Vygotsky (2007) explains that the dynamics of the intemalisation process take place over long periods of time, through successive changes and transitions where social and individual psychological functions appear related or overlapping, to finally become individual. An example of this dynamic is the formation of a child's language, which develops from social speech, through a transitional stage of egocentric speech, to the intrapsychological level of inner speech.

Therefore, after describing the concepts involved in the process of human psychological development from the socio-historical perspective, we can understand that the theory points out that some aspects of the individual's development originate from their interactions in a cultural context. Mediation is an important element in helping students to internalise new concepts, leading them to become more autonomous in the development of their learning until they reach the level of self-regulation. Thus, for students to reach the level of intemalisation through mediation, they need some mechanism to understand and monitor this process. Therefore, in the next section we'll look at such a mechanism.

2.5 Learning mediation categories

One of the current problems for socio-historical researchers is to follow and understand the process of mediation until the individual reaches intemalisation.

Some researchers have identified the existence of a mechanism that could identify an intemalised sign and therefore accompany the subject's cognitive development (DIAZ et. al., 1993; GALLIMORE and THARP, 1996; WERTSCH, 1999).

For Diaz et. al. (1993) "[...] the common denominator of these transformations or evolutionary changes is the diminishing power of the immediate contingencies of the environment and the growing role of the reformulation of projects and objectives in the regulation of behaviour and cognitive activity" (p. 153) and what he called self-regulation. Self-regulation capacities develop within the context of social interaction between individuals who are more and less experienced in the process. It would be through this context of social interactions that the process of intemalisation could be verified.

Thus, from an educational point of view, mediation actions are important during the development of the process of social interaction between individuals, not only for the cognitive development of the student, but also for the development of their autonomy. Virtual teaching-learning environments are no different. These environments have mechanisms that allow for both teaching and flexible learning, independent of the space and time in which they can be developed.

Furthermore, learning, mediation and communication are closely linked concepts that represent a key point for the quality of distance learning. The teacher has an important role to play in his or her mediation activity, seeking to make adjustments so that the student can develop a better performance, according to the student's individual needs.

The regulation process proposed by Diaz et. al. (1993) and adapted by Passerino (2005) is carried out in levels that start with control, go through self-control and reach self-regulation. These levels of learning mediation make it possible to monitor or check the progress of students' learning towards self-regulation.

Control is external to the subject, carried out by the more experienced subject and can take two forms: direct or indirect. **Direct control takes** the form of orders, directives and directive questions. **Indirect control, on** the other hand, takes the form of perceptual, conceptual and procedural questions, culminating in physical distancing (the more experienced subject leaves the less experienced subject to observe). The gradual physical withdrawal of the more experienced subject in the learning mediation process aims to lead the student to enter the level of self-control.

Diaz et al. (1993) consider **self-control** as the student carrying out an expected action in obedience to an internalised tutor. In other words, the figure of the more experienced subject, who was real and external to the control process, is now internal, but still exists as another subject in the student's cognitive apparatus.

Self-regulation cannot be observed directly, as it happens intrinsically to the subject, but the subject is considered to be in the self-regulation category when he or she organises, plans and executes the action without the intervention of any external mediator. Thus, self-regulation is the action plan conceived by the subject who becomes their own tutor.

The central differentiation between self-control and self-regulation does not lie in the intemalisation of the tutor's orders and directives, but in the emerging capacity to plan and define one's own objectives, functionally organising one's conduct towards them and adapting it according to the context. Figure 1 presents an integrated view of the categories of mediation, illustrating the possibilities of pedagogical intervention that take place in an educational process.

The original studies by Dias et al. (1993) obtained the mediation categories from empirical research based on observing mothers and children solving specific tasks. The studies identified two categories: self-control and self-regulation. In the work by Passerino (2005), it was considered important to include the category of control, since not all the subjects observed could start from self-control. In addition to this adaptation to the original model by Dias et. al. (1993), two dimensions were included: direct and indirect control in each category, with the aim of improving the process of monitoring the mediation phenomenon, which was not included in the original study. These dimensions were included because of the support that the mediator offers their students (PASSERINO, 2005).

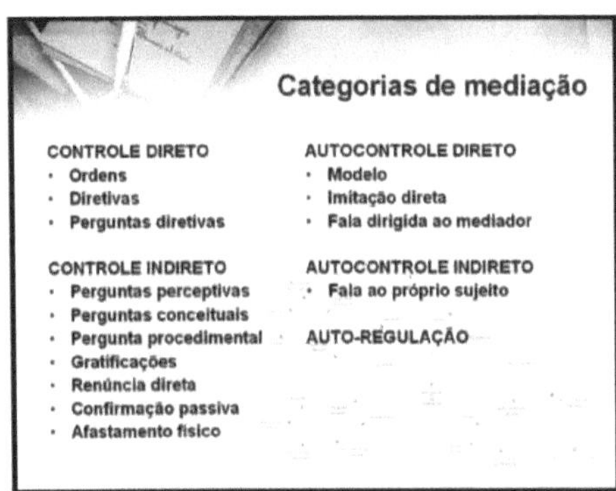

Figure 1 - Mediation categories
Source: author

Based on the work carried out by Dias et. al. (1993) and Passerino (2005), a study was carried out to identify these categories of mediation in data from the textual interactions of participants in virtual teaching-learning environments. This study, carried out by researchers specialising in the field of education, aimed to identify signs of mediation in *online* environments by identifying textual markers that would make it possible to identify mediation categories. Therefore, the next section presents possibilities for applying mediation categories to data from virtual environments.

3. 3Pedagogical mediation   in VLEAs

The work of Koch et al. (2009) points out that evidence of mediation in distance learning environments is of great importance in the development of educational processes. The data collected from empirical research shows that from a pedagogical point of view, the role of the mediator and their adjustment to the help offered within the identified categories of control, self-control and self-regulation are important for student autonomy and the appropriation of knowledge.

The study's basic hypothesis is that the categories of control, self-control and self-regulation can be applied productively to analyse and adjust the mediation processes that take place in distance learning environments, because as well as helping to understand mediation processes, they propose that these categories can serve as a basis for building tools for virtual environments that support both teachers and students in achieving productive mediation processes from a pedagogical point of view.

The paper points out that in an educational process, mediation "adjustments" are often made as the work progresses and their monitoring and regulation is only possible by "reading" the intersubjective context that is established in social interaction and that contains not only linguistic elements, but also extralinguistic elements that provide knowledge about the definition of the shared situation. He describes that this definition of a shared situation comes from what he calls the environment. An environment is the element that circumscribes a linguistic activity. Environments are not only present, but they are especially guiding discourse, as a kind of backdrop that effectively contributes to the realisation of discourse and its meaning and sense-making. Environments are of four types: situation, region, context and universe of discourse (Andrade 1998 apud Koch et. al. 2009).

In this study, the situations highlighted are specific interactions within the environment using communication tools such as mail, portfolio, diary and forum. The region is the space within which a sign functions in certain systems of signification. The region is delimited by the distance education environment. The context is related to the whole reality surrounding a verbal activity and *is* subdivided into three: idiomatic, verbal and extra-verbal. The first is the language used in speech or communication. The verbal context is the discourse itself and the extra-verbal context is the set of extralinguistic circumstances that are perceived or known by the interlocutors and is everything that physically or culturally surrounds the act of enunciation.

In most of today's virtual teaching-learning environments, the context of interaction represented that allows a definition of a shared situation to be generated is merely the verbal context, leaving extra-verbal contexts out of perception, justifies the study presented. This presents two problems:

- All interaction is centred mainly on written discourse, i.e. linguistic elements, with the extra-linguistic context restricted to a few verbal markers such as punctuation marks, blank spaces, paragraphs, *emoticons,* etc.

- There is also a time lag due to the flexibility of space and time that distance learning allows. This lag doesn't always favour mediation, as it generates a certain amount of anguish and a sense of failure.

The study highlights the need for mediation to permeate the entire environment. The search for evidence is only possible if it is centred on the actors, relating all the tools used so that space-time flexibility and extra-verbal contexts do not go unnoticed.

Based on the study presented here, we move on to a contextualisation of pedagogical mediation and the relationship between the categories of mediation and the work carried out in this doctoral thesis.

3.4 Mediation and learning

The study by Koch et. al. (2009) is part of this research and is also a stage in the MEDIATEC project already explained in this thesis. Based on the results obtained by the researchers, the categories of mediation were identified in texts from the interactions of participants in virtual teaching-learning environment courses.

The categorisation of the texts aimed to identify the levels of mediation of the interactions carried out in different tools of the virtual teaching-learning environment. Thus, we identified instances of interactions referring to categories of mediation in relation to student interactions and interactions carried out by the teacher or tutor of the course.

In this study, we consider the process of personalisation based on three processes that are active in the student's cognitive development: control, self-control and self-regulation. It should be emphasised that the level of mediation of the student's learning does not necessarily start in the control category, where the mediator needs to be more present and active in pedagogical intervention actions. Depending on the student's level of knowledge and learning development, their interactions can be identified at any level of mediation.

Below we will describe each category of mediation in more detail, as well as its subcategories, presenting examples of textual markers that allow us to identify the level of mediation of each textual interaction between students and mediators during the teaching-learning process.

We begin by describing the control category from the point of view of the mediator and the student. From the mediator's point of view, the student's process of appropriating concepts must be accompanied by more frequent pedagogical interventions. As we saw in previous sections, this control can be direct or indirect.

Direct control takes the form of orders, directives or direct questions, as we'll see below:

- **Orders:** in mediation by direct control through orders, the mediator guides the student's learning process through imperative actions. Examples of these imperative actions are: "do it like this", "write a text" and "post contributions on the forum".

- **Directives:** the action of mediation by direct control in the form of directives is carried out by softer orders aimed at guiding the student in the development of their learning. When communicating with the student, the mediator uses the plural to soften their imperative action. Examples of these interventions are: "let's do it this way" and "it would be interesting if you posted your contribution on the forum".

- **Direct questions:** in direct control mediation in the form of direct questions, the mediator uses questions containing implicit orders. For example: "Can you put the text on the Wiki?".

In the case of intervention by indirect control, the mediator can ask perceptive questions, conceptual questions, gratifications, direct renunciation, passive confirmation or physical withdrawal. Below is a description of each of these subcategories of indirect control:

- **Perceptual questions:** these are questions addressed to the student whose answers are related to their perceptual field. Examples of this type of question are: "What colour is this shape?" or "Is the button rectangular or round?".

- **Conceptual questions:** this type of question encompasses those that cannot be answered by immediate perception and require reflection. Examples: "What next?" or "What is the difference between the two objects shown?".

- **Procedural question:** a question about a procedure to be carried out. "Tell me how to post the message on the forum?".

- **Gratifications:** when the mediator uses praise during the development of the student's learning. Examples: "that's right, keep it up" or "congratulations, you're doing well".

- **Direct resignation:** in this type of intervention, the mediator already aims to withdraw, leaving it up to the student to solve a particular problem. To do this, they use actions such as: "now do it yourself".

- **Passive confirmation:** the mediator responds to questions addressed to him by confirming or not confirming the actions carried out by the student. Confirmations can take the form of: "yes, that's correct" or "no, try another way".

- **Physical distancing:** the mediator leaves the student alone to carry out the task, only observing and intervening when necessary. This type of pedagogical intervention is already related to the self-control category.

From the student's point of view, the internalisation of concepts through control actions can also be direct or indirect. Direct control actions included imitation and verbal response. As described below:

- **Verbal response:** directed by the student to the mediator's questions. Examples: "I confirm receipt of the e-mail" or "ok, I can do it in the afternoon".

- **Request for help:** direct request for help from the student to the mediator. Example: "I couldn't open the document".

As for indirect control actions, there were instances of guided questions and verbal responses. Below is a description of each subcategory observed:

• **Guided questions:** questions posed to the mediator or other colleagues on issues covered by the subject. They can be procedural or conceptual.

Therefore, we intend to apply a computer technology in virtual teaching-learning environments to map and monitor the levels of mediation of participants' interactions during an *online* course. The next section describes the concepts involved in virtual teaching-learning environments, seeking to contextualise the application of this research work.

CHAPTER 4

VIRTUAL TEACHING AND LEARNING ENVIRONMENTS

Andrade and Vicari (2003) point out that the Web has become a space to support the development of distance learning courses supported by a variety of *online* computerised environments aimed at educational activities. In this educational modality, the teacher plays the role of mediator during the learning process, with the student at the centre of the process. A special feature of this type of education is that the teacher is not a mere transmitter of knowledge, but rather a moderator who prepares a space for dialogue and interaction for the students. Students are no longer passive recipients, but builders and socialisers of knowledge.

Kenski (2007) reports that the way we teach and learn has changed a lot since ICT began to expand in society. Both students and teachers have intensive contact with media equipment in their daily lives, such as televisions, computers and the Internet. In this way, they remember the information and experiences they had through their interactions with this equipment. As such, the author emphasises that these mediations make it possible to understand that teaching and learning activities are not exclusively linked to face-to-face environments.

According to Torres and Fialho (2009), the use of ICT has introduced more interactive environments into the distance learning modality, enabling collaborative *online* learning, where interactive technologies have been added to previous media such as radio, television, audio tapes, telephone, etc. For Teles (2009), web-based teaching has great potential for collaborative pedagogical models, which are demonstrated by the following characteristics:

- group-to-group communication, which allows students to communicate directly with other classmates via the *online* environment',

- temporal independence, since access can be made at any time, which allows students the time they need to reflect on and analyse the topics posted in discussions in the *online* environment',

- spatial independence, as the online environment can be accessed from anywhere, as long as you have an Internet connection;

- interaction between participants via computer-mediated communication that requires the student to organise ideas through writing.

According to Nunes (2009), we are currently experiencing a new wave that involves the mastery of a communicative technology based on telematics (a combination of computer and telecommunications technologies), which is articulated through virtual organisation supported by networks. ICT and its educational applications can create conditions for more interactive learning, through non-linear paths in which students can determine their pace, speed and route.

Araújo and Marquesi (2009) add to this by saying that ICT provides the potential for new learning experiences, which should not be overlooked by teachers, as they make it possible to devise strategies so that students can achieve their learning objectives. In addition, teachers should bear in mind that the opportunity for different learning experiences for students can only be achieved if they are well guided by the teacher. To this end, teaching approaches appropriate to the virtual context should be adopted. Tiffin and Rajasingham apud Araújo and Marquesi (2009) present two epistemic approaches to teaching. A traditional teaching and learning approach based on the following principles:

- teacher-based process;

- through exposure to knowledge;

- closed learning;

- authoritarian attitude;

- individual study;

- face-to-face interaction;

- classroom-based action;

- learning administered by the teacher.

And an episteme based on the context of virtuality[5] , which is supported by the principles listed below:

- student-based process;

- through reflection and problem-solving;

- flexible learning;

- democratic stance;

- group study (co-operation and collaboration);

- Internet interaction;

- action centred on hypertextuality;

  computer-mediated learning.

Valente apud Silva e Silva (2009) presents other possibilities for distance learning with the help of the Web: (a) the *broadcasf* approach, (b) the traditional virtual school; and (c) being together virtually. The authors emphasise that these approaches are used according to the degree to which they contribute to the process of building knowledge.

In the *broadcast* approach, interaction between student and teacher is practically non-existent, since the content is made available to the student on a website, so that they can access it at any time, but the teacher does not monitor the development of learning. This approach does not guarantee the construction of knowledge and consequently does not guarantee educational quality, although it does serve the purpose of disseminating information to a large number of students.

In the traditional virtual school approach, the process is centred on the teacher, who acts as the holder of the information that will be passed on to the student. Interaction between student and teacher takes place via the Internet with the aim of checking that the student has processed the information, which can be achieved by carrying out assessments. The virtualisation of the traditional school is of higher quality than the *broadcast* approach, but it does not guarantee

---

[5] We use the term based on the thinking of Lévy (1996), who discusses virtuality as a way of potentiating the teaching-learning strategies provided by the teacher in the teaching modality supported by the Internet.

sufficient conditions for the construction of knowledge. Valente apud Silva e Silva (2009) points out that the existence of some interaction between student and teacher means that fewer students can be catered for than with the *broadcast approach*.

The last approach, being together virtually, involves a more intense interaction between student and teacher, as there is more constant counselling and monitoring of student activities in order to understand and propose challenges aimed at developing learning. In this approach, the teacher must act by providing opportunities to build knowledge and the student must reflect and try to put their ideas into action in order to solve problems. Students' doubts should be referred to the teacher and the group, so that a cycle of co-operation develops during the learning construction process.

It should be emphasised that learning takes place independently of the existence of teaching. In contemporary society, learning has become a necessity for adapting to society. Learning has therefore become essential (ARAÚJO and MARQUESI, 2009). As such, *online* educational environments have emerged as a way of expanding possibilities, seeking autonomous, flexible learning that favours interaction between teacher and student.

4.1 Technologies in *online* educational environments

According to Peters (2001), one factor that influences the communicational distance between teacher and student is the structure of the teaching programme, which, according to the author, the more pre-programmed and compulsorily prescribed to students, the greater the communicational distance between individuals. For Franco et al. (2003), *online* education *is* an educational modality that adopts strategies aimed not only at overcoming the physical distance between educators and students, but also at bringing them closer together.

Moran (2004) adds that *online* education is a type of teaching that aims to help participants balance their personal skills with participation in face-to-face and virtual groups, which allows the individual to advance quickly through the exchange of experiences, doubts and results. These reflections are in line with the concept of transactional distance proposed by Moore (2007), which does not consider physical distance, but rather the communicative distance between teacher and student. According to this concept, transactional distance is inversely proportional to the intensity of communication between the participants in the teaching-learning process.

Teles (2009) points out that in the last three decades the increase in computer-mediated human communication for educational purposes has led to a proliferation of technologies for the purpose of offering *online* educational environments. Litto (2009) emphasises that with the development of web technologies, universities and commercial companies have begun to exploit their potential in the form of educational environments.

Complementing this, Nunes (2009) states that the main innovation in the field of education in recent decades has been the creation, implementation and improvement of a new generation of distance learning systems, which has opened up possibilities for promoting educational opportunities for large contingents of the population. Not only in terms of quantity, but also with a strong sense of quality and flexibility.

Araújo and Marquesi (2009) cite other relevant factors regarding technologies in *online* educational environments, which are: knowledge, competence and ability to determine and use digital technologies in a given teaching strategy. These factors make it possible to identify the potential of the *online* educational environment for a particular strategy to be adopted by the teacher. Students must have knowledge of the technology, as well as the ability to search for information and construct meanings that can be useful.

Litto (2009) emphasises that web-based courses are growing a lot due to the acceleration of innovations that also extend to the area of communication. The author states that it is impossible to say that there is a dominant standard of technology for *online* educational environments, as this is still new for environments as varied as schools and corporations. It should also be noted that the use of technologies such as mobile phones or *iPods has* not caused any paradigmatic revolution, but they are changing the way institutions use technologies in teaching and learning. In the past, institutions determined the technologies that students would use; today, leaders are trying to adapt to the technologies that students already have at hand.

Franco and Costa (2005) emphasise interactivity as an important aspect in *online* educational environments, *as* well as collaborative learning. The authors mention that the preparation of pedagogically valid self-learning materials and teacher mediation are key elements in the process. Pedagogical materials should contain challenging activities that encourage dialogue. The mediator must interact with the student to reinforce their self-learning. For the authors, collaborative learning is strengthened by the creation of learning networks formed by the sharing of ideas through tools such as email, discussion lists, chat, etc. The tools provided by ICT in the educational sphere enable many possibilities for interaction, favouring the dialogic process that is so necessary in *online* educational environments.

Next, we present concepts about virtual teaching and learning environments, since the focus of the studies in this thesis is on the process of pedagogical mediation, exercised by teachers and students in the development of *online* education with the support of these environments.

4.2 Virtual Teaching-Learning Environments

Silva and Silva (2009) put forward the idea that AVEAs are spaces that allow interactions to be established at various levels, where the roles of teachers and students can be re-dimensioned and hierarchisation can be replaced by mediation between individuals. These environments enable different ways of thinking about learning due to the characteristics of time and space provided by these platforms and where collaboration must replace competition due to the concept of networking in the context of virtuality. The authors emphasise that such environments are conducive to the development of student autonomy and creativity, which is hardly ever encouraged in traditional educational environments. In addition, Pontes apud Silva e Silva (2009) points out that student autonomy is an indispensable condition and that this is characterised by the development of the ability to research, organise and think critically and independently.

Coutinho (2009) states that the growing demand for technologies to support teachers in their tasks of managing *online* educational environments led to the emergence of virtual teaching-learning environments. These systems were designed to administer courses by registering users and managing information on the development of *online* activities. According to Vaz (2009) the focus of VLEAs is the student, where the main objectives are to manage students and their learning activities. These systems help to administer and monitor relationships between users through profile information, history, user roles/functions and preferences.

We need to understand the concept of a virtual learning environment for teaching and learning, highlighting its characteristics and potential as an educational platform for distance learning. According to Almeida (2003), VLEAs are computer systems that use the Internet as a means of accessing educational activities mediated by ICT. These environments allow for the integration of different media, languages and resources. In addition to enhancing interaction between participants and objects of knowledge, they seek to achieve certain previously set objectives. We must realise that it is not the tools available in these environments, nor even their structure that will guarantee learning, but rather the way in which these resources are used to build collective knowledge based on the interactions of individuals, guided by prior planning.

VLEAs were designed with the intention of reducing the transactional distance in *online* education, aiming to improve the quality of this educational process, emphasising the importance of structuring this platform so that it allows constant interactivity between participants. The first initiatives to build virtual learning environments took place in the mid-1990s, influenced by changes on the Internet, such as: popularisation and growth in the use of the network with the incorporation of business organisations; and the emergence of the Web, through the use of the browser and its standardisation as an interface for accessing content (FRANCO et. al., 2003).

According to Franco et al. (2003), VLEAs are more effective if they include mediation activities. According to Salmon (2000), the pedagogical characteristics of virtual teaching-learning environments occur in four technological scenarios, which the author considers to play a role in guiding (mediating) students during the development of their work, as well as in assessment processes.

- The first scenario is based on an emphasis on content, the pedagogical correspondent of which is the pure transmission of information from the teacher to the student;

- The second scenario is related to the need for continuous learning, based on the need to develop new work fronts in certain fields, where the focus is on developing skills, competences and autonomous learning;

- The third scenario is related to student mobility, caused by the dynamism of today's society, which demands independence of time and place, based on the pedagogy of learning styles;

- The fourth scenario has to do with learning communities, seeking to favour collaboration and cooperation in the construction of individual and collective knowledge.

According to Salmon (2000), the intervention of moderators in VLEAs is essential for the engagement of participants, avoiding disappointment and frustration of expectations, which is one of the problems identified in distance education. Moderation must take into account issues such as access and participation, learning styles, the nature of the course and the participants. Franco et. al. (2003) emphasises that moderation is a relevant aspect that can influence both positively and negatively the development of an educational project based on AVEA.

Querte (2003) emphasises that interactive virtual environments are important; however, the teacher's preparation and support in their interventions is more important for collaboration and the development of learning to occur. The dialogic interaction strategies used by mediating teachers are important for encouraging students to carry out different forms of interaction in the environment, always seeking to develop knowledge (FAVERO, 2008).

Okada (2003) adds that another problem with the educational modality based on virtual environments is the price of the temporal and spatial flexibility provided, since it can overload the teacher when organising the large flow of information generated, requiring a significant amount of time. In this way, the author emphasises the importance of pedagogical mediation in VLEAs as a way of encouraging collaboration.

Franco and Costa (2005) complement the importance of pedagogical mediation by stating that the teacher must act as a facilitator during the process of constructing learning in virtual environments, acting in the selection of content, establishing logical sequences, identifying materials and sources, moderating shared spaces, monitoring production spaces, as well as developing strategies for setting up a collaborative learning environment.Among the ideas reported in the previous paragraphs, there is a clear concern with pedagogical mediation as an important factor in the development of collaborative activities and the construction of knowledge based on the interactions of individuals in the AVEA. The authors emphasise the importance of the teacher and their role in moderating activities, acting as a mediator during the teaching-learning process in these environments.

4.3 Current problems and future trends in AVEA

Studies such as Vieira's (2011) seek to analyse the perception of teachers and tutors of distance learning courses. The study collected relevant data on the use of information and communication technologies in the learning process of distance learning students. The author emphasises that information and communication technologies are guiding elements in learning, enhancing integration between participants and the desired knowledge. Rodrigues (2011) addresses the problem of pedagogical practices in distance learning, highlighting the need for teachers to take ownership of all the technologies made available by *online* environments.

Messa (2010) points out that virtual teaching-learning environments are increasingly being used in academic environments, highlighting the importance of understanding the concepts and structures of these environments. As well as how these environments can support the teaching-learning process. According to the author, it is important that AVEAs are equipped with media such as video, text, audio, graphics and animations, with the aim of developing students' skills and concept formation, as well as enabling numerous learning modalities.

Coutinho (2009) reports that Web 2.0 has brought about not only technological but also social changes, since the Internet has gone from being a repository for searching for information to a place where people also write and produce content. Stephen Downes apud Coutinho (2009) points out that the Internet has become a place where information circulates through documents that are created, shared, modified and revised. As a result, educators realised that in order to meet the needs of creating and organising courses, they couldn't just rely on the functions that an AVEA should contain, but it was also necessary to think about the pedagogical project aligned with the organisation's educational objectives.

The author points out that the future increasingly points towards flexibility and modularisation of courses, where the trend is for various components to be created and added to the structure of VLEs in the form of modules. In this way, AVEAs will have to be prepared for the coupling of various modules that will be available according to the pedagogical needs of each course.

Coutinho (2009) emphasises that many education professionals have started to use a combination of applications available on the web, dispensing with the use of an AVEA. These professionals consider that learning is also informal, which is why they use other tools outside the *online* learning environment. In order to learn, one must consider not only the content available on the AVEA, but also the possibility of searching the Internet via blogs, portals, social networks, wiki, etc.

Feldstein and Masson apud Coutinho (2009) defend the flexibility of VLEAs through the extension of functionalities by means of attachable modules, as long as they are based on Web 2.0 resources. Valente and Mattar apud Coutinho (2009) state that various Web 2.0 resources are being used in education and that they are inserted into the architecture of a course according to the pedagogical project and educational strategy planned.

As for future prospects, Fialho and Torres (2009) state that a generation of VLEA is emerging on the current world stage characterised by the adoption of artificial intelligence and virtual reality concepts. In this way, there is an environment where the student interacts directly with the machine, which is responsible for managing learning using AI techniques; as well as environments that will enable immersive learning using virtual reality principles. The authors emphasise that VLEAs will be influenced by the concept of the semantic web, which will make it possible to obtain more precise information due to the use of intelligent agents to refine information search engines on the web.

Antunes et al. apud Fialho and Torres (2009) comment on a project that aims to integrate Moodle with Second Life, in an environment called Sloodle. This project maintains similarities with the basic functionalities of an AVEA, but includes tools for building objects in three dimensions. It also includes tools that allow interaction with external sites. Valente and Mattar apud Coutinho (2009) consider this environment to be an excellent platform for promoting flexible *online* education, despite the fact that there are still a number of limitations such as: time taken to learn the tool, difficulties in installing and configuring the system, connection problems, hardware requirements, etc. Having explained the current problems and future trends in VLEAs, we move on to contextualise this research in relation to *online* educational environments.

4.4 Online educational environments

In this thesis we emphasise the expansion of the pedagogical mediation process in virtual teaching-learning environments from a socio-historical perspective. With the study proposed here, we intend to specify, develop and implement a support technology for surveying signs of mediation during the development of student learning in AVEAs.

Based on the cognitive categories outlined in the work of Dias et al. (1993) and later in Passerino (2005), we intend to apply an extension to the Moodle virtual teaching-learning environment, in order to verify possibilities for monitoring the development of student learning mediations, based on interactions carried out in the environment. This will help teachers and tutors with elements for their pedagogical intervention strategies in the virtual teaching-learning environment, according to Araújo and Marquesi (2009).

In this research, *it is* important that there is intense interaction between the participants in the *online* course, so that we can validate the results of the application of the system for mapping signs of pedagogical mediation in the virtual teaching-learning environment. To do this, we will rely on a course that is structured according to the virtual being together approach outlined by Valente apud Silva e Silva (2009). Since the teacher must be an advisor and propose challenges aimed at the constant development of student learning. Students' questions and contributions should be addressed to the group so that we can identify signs of mediation throughout the course.

Through the study proposed here, we intend to reinforce the ideas of Alves (2011), Messa (2010) and Litto (2009) about the growth of web-based courses in line with the innovations introduced in *online* educational environments. In this case, through the contributions that this study will make to the development of *online* courses with VLE support.

In addition, we rely on what Coutinho (2009) points out regarding the tendency for new components to be created and added to the structure of current VLEs in the form of modules. As well as making the environment more flexible, as explained by Feldstein and Masson apud Coutinho (2009). In this way, technologies are developed to support the pedagogical strategies of *online* courses. In this case, the proposal of this thesis is to design a technology to support mediation.

The technological basis for surveying evidence of mediation in the virtual teaching-learning environment will be text mining technology. This thesis therefore includes a description of the concepts and terminology involved in text mining. It is also necessary to contextualise text mining technology with the research carried out in this thesis. The next chapter provides a general explanation of text mining and its application in this research project.

CHAPTER 5

TEXT MINING AND MEDIATION

In this chapter we will discuss the subject of text mining, as it forms the technological basis for the application of the study involved in the research presented in this thesis. To carry out the research, we used algorithms to classify levels of mediation, based on textual data resulting from the interactions of participants in an *online* course in a virtual teaching-learning environment.

The following sections present the theoretical background and state of the art in research into the use of text mining techniques in educational applications to obtain the usual knowledge for teachers to make pedagogical decisions in *online* educational environments. We therefore begin the next section with an explanation of the concepts involved in text mining.

5.1 Text mining

The term text mining is related to a process for discovering patterns of knowledge that are of interest to the user from collections of data in textual format, as described by Tan (1999). To do this, Bisisoly (2008) emphasises that a technique is used whose aim is to extract new and useful information from a set of words that make up a text.

The discovery of patterns in textual information is at the heart of the research carried out in this thesis, since the search for evidence of mediation in the virtual environment is based on an analysis of the interactions between participants in the *online* course. As a large part of these interactions in the virtual environment are stored in textual form, we adopted the use of text mining in this research approach.

Moty (2005) emphasises that a text document can be seen from various perspectives. As a structured object, since from a linguistic analysis, a textual document can be considered a rich collection of syntactic and semantic structures, although this structure is implicit. The author also mentions that typographical elements such as punctuation marks, numbering and special characters, associated with formatting artifices for the *layout of* textual presentation, such as: spaces, emphases, underlining, tables and other elements, help to identify important textual components such as titles, paragraphs, authors' names, headings, footnotes, etc. In addition, some text editors include metadata in the structure of the text to help identify it. This type of text is considered semi-structured because it contains information about the content or structure of the document.

However, data in textual form does not follow a well-defined logical structure. This gives rise to a research problem for text mining researchers: how to extract patterns from a set of texts, producing new information? This gave rise to the area of text knowledge discovery, as explained by Feldman (1995; 1998).

For Rajman (1998), due to the form of a textual document, specific techniques are needed for knowledge extraction, since textual documents also store useful information and potential for generating knowledge.

Therefore, in a text mining process, analytical functions from areas such as natural language processing and information retrieval are applied (MOTY, 2005). In this sense, researchers use text mining tools to:

- Extraction of relevant information from a document, such as entities that represent characteristics of a text, through techniques such as natural language processing and information retrieval.

- Searching for trends and relationships between people, places, organisations, etc., by aggregating and comparing information extracted from a set of documents.

- Classifying and organising documents according to their content.

- Document retrieval based on various types of information about the document's content.

- Aggregation of documents according to their content.

For Sumathi and Sivanandam (2006), text mining should not be confused with the technologies of Internet search engine tools or the specific capabilities of database management systems. The authors report that there are three main technological components involved in text mining:

- **Information retrieval:** this is the basis of the activity, as it involves extracting relevant and important records from text sources for further processing.

- **Information processing:** extracting patterns from a set of data retrieved in the information retrieval stage. This phase aims to sort or classify unstructured data found in texts.

- **Information integration:** combining computer-processed information into an output format appropriate to the human cognitive process, using visualisation techniques.

> Text collections are the focus of the text mining task and are characterised by any group of text-based documents. These collections can be static or dynamic. Static text collections are those that are immutable from the moment they are created, i.e. they do not change over time; whereas dynamic text collections undergo constant changes through the updating or insertion of new documents. The performance of text mining tasks is affected according to the nature of the document collection.

Below we describe the stages of a text mining process. Feldman and Sanger (2007) explain that text mining can be divided into three main stages: data preparation, mining and post-processing.

It should be noted that text mining systems do not usually apply their algorithms to collections of documents that have not been previously prepared. Therefore, greater emphasis is placed on the pre-processing phase of operations. Sumathi and Sivanandam (2006) report that pre-processing operations are made up of a variety of techniques taken and adapted from the field of information retrieval and computational linguistics, which meet the initial need to transform unstructured data, which is the original form of textual documents, into an intermediate data format suitable for the algorithms to be used in the analysis task.

Text pre-processing operations for the mining process must take into account the different elements of natural language contained in a document in order to transform an implicit and irregular structure into a structured and explicit representation. This is an important aspect for any knowledge extraction task in textual databases, since a mining algorithm works on the basis of the representation of a document's characteristics. The characteristics most commonly explored by mining algorithms in a textual document are (MOTY, 2005):

- **Characters:** these form a level of individual text components. They are a minimal element, made up of letters, numbers, spaces and special characters. They form the building block of high-level semantic structures such as words, terms and concepts. Textual representations at the character level can include sets of all the characters in a document or just a few according to filtering criteria. Textual representations using character *bag* approaches without positional information are not very useful for text mining purposes, while positional character set representations have some usefulness for the mining process in techniques such as *bigram* and *trigram*. Although they do have some utility, approaches based on character representations don't generate much knowledge about real-world text documents.

- **Words:** this is a basic semantic level of document representation. As such, textual characteristics at the word level are often considered a native representation of documents. A single word is considered a linguistic *token* at most. In sentences, multi-word expressions, compound words do not constitute a single *token.* The usefulness of a textual representation at word level is the use of techniques for extracting *stopwords*[6] , symbolic characters and insignificant numbers.

- **Terms:** these can be simple words, compound words or phrases, extracted directly from a document's *corpus* using term extraction methodologies. For example, let's take the following sentence: "the student used his new notebook on his first day at school". A list of terms representing the

---

[6] *Stopwords* are words without much linguistic meaning for the purposes of analysing and extracting information.

previous sentence could consist of just words, such as: student, notebook, new, day, class; or a set of expressions found in the text, as in: "student used", "new notebook", "first day of class". Term extraction methodologies can convert texts in a document into a series of standardised terms.

- **Concepts:** these are characteristics generated for a document, either manually or through statistics, rules or categorisation methodologies. Concept-level features can be generated manually for documents, but are more commonly extracted by complex pre-processing routines that identify words, expressions, entire clauses, or larger statistical units that are classified as specific concept identifiers.

Of the four characteristics exploited by text mining algorithms, terms and concepts are the most relevant and expressive at a document semantic analysis level. As such, they bring many advantages to the representation of documents for text mining purposes. The representation of documents by concepts is more widely applicable in the process of text mining, as it can clearly extract synonymy and polysemy in a given document, supporting the definition of sophisticated hierarchies of concepts that serve as the basis for the creation of ontologies and knowledge bases.

The pre-processing stage involves cleaning the data in order to facilitate the next stages of the process. Data cleaning is the removal of words considered unnecessary for understanding the text. The pre-processing stage is organised into three stages: correcting the spelling of the text, removing *stopwords* and *stemming*. Although the pre-processing process necessarily follows these three stages in the order they have been mentioned, it is not compulsory to carry out all of them; some can be omitted (KOWALSKI, 1997) (KORFHAGE, 1997) (KRAAIJ, 1996).

- **Spelling correction: the** aim is to eliminate possible spelling mistakes in the text to be analysed. A spell checker is used to carry out this stage, checking the text.

- **Removal of *stopwords'. The*** purpose of this stage is to remove repetitive words from a text or words which, if removed, will not make much difference to the understanding of the text. This stage removes digits, punctuation marks, converts characters to lower case and removes prepositions, articles and conjunctions. The removal of *stopwords* aims to reduce the length of texts by 40 to 50 per cent (SALTON, 1983).

- ***Stemming:*** words in a text are made up of many morphological variations. Therefore, this stage seeks to eliminate this variation using lemmatisation techniques (KRAAIJ, 1996).

The next stage is the preparation and selection of the data in order to find the most representative terms obtained in the pre-processing stage. This stage is necessary because in a large collection of documents the vocabulary of terms generated can be unmanageable by text mining techniques. Therefore, it is important to select the most relevant terms that can identify the texts, and to do this you need to use a number of techniques:

- **Relevance calculation:** uses a measure to calculate the relevance of a given term in relation to the text. The most commonly used methods are based on the frequency of terms in a text, such as: absolute frequency, relative frequency and document frequency (RIJSBERGEN, 1979).

Absolute frequency, or *term* frequency (TF), refers to the number of times a term occurs in a document. It is the simplest measure of weight, but disregards the distinction between terms that occur in a few documents and terms with a high occurrence in many documents. A

absolute frequency does not take into account the number of terms in a text, for example, a certain term that occurs many times in a small document may

be less relevant than a term that occurs only a few times in a long document, or vice versa. Relative frequency aims to get around this problem by taking into account the number of terms in a document (size) and normalising weights accordingly. Figure 2 shows an expression for calculating the relative frequency of a term in a text by dividing the frequency of the term *(Tf)* by the total number of terms found in the text (TV).

$$Frelx = \frac{Tf(x)}{N}$$

Figure 2 - Calculating the relative frequency of a term Source: Feldman and Sanger (2007)

- Another important measure is *document* frequency, which seeks to get round the problem of absolute frequency, where the number of documents in which a term appears is not taken into account. By combining the techniques of term frequency and document frequency, it is possible to increase the importance of terms that appear in a few documents and decrease the importance of terms that appear in many documents. According to Salton (1983), this is because terms with a low document frequency are generally more discriminating.

- Kowalski (1997) highlights an important observation about the identification of frequency weights in document collections, which is the temporality of the frequencies, since the set can vary with the inclusion of new documents, or by changing the content of documents in the collection.

Document representation: based on the relevance of terms in a text, data can be coded to choose a document representation model, which will be used by statistical methods to extract knowledge. The document representation model is chosen according to some criterion that reflects interesting properties of the data in relation to the modelling purpose. In addition, the choice of model must take into account a format that can be understood by the text mining techniques used.

Bekkerman et. al. (2003) explains that in text mining research, documents are coded in the form of a *bag of words,* where each document is represented by a vector containing the terms contained in its content. This technique simplifies the representation of the information contained in documents, but does not express a faithful description of their content. Two widely used models are Boolean and vector space.

The Boolean model uses a vector of logical values "0" or "1", which indicate the absence or presence of a particular term in the document. The main advantage of this approach is that it is simple and saves space for storing information. However, this representation is limited and the vector space model tries to get round this problem by associating weights to the terms found in a document, which allows partial similarities between terms and documents to be discovered.

Selecting attributes: considering all the terms present in a document is often not feasible, as it can jeopardise the performance of the mining algorithms. Therefore, selecting the most relevant attributes is an important step, as it aims to eliminate terms that are not representative of the domain of knowledge to be extracted. An attribute can represent one or more terms contained in a document. However, the selection of attributes must be a careful task and supported by appropriate tools. To this end, techniques have been developed to aid the attribute selection process, such as: filtering by weight, selection by latent semantic indexing, selection by natural language, etc.

Weight-based filtering consists of eliminating terms whose frequency weight is lower than a predetermined standard set by the user. Even so, the dimensionality of the filtered terms can still be high, so a selection of the n most relevant terms should be applied by truncating them to establish a maximum number of terms that can represent and characterise the information in a document (YANG and PEDERSON, 1997).

The latent semantic indexing selection method aims to reduce the dimensions of the vector space model used. This is achieved by analysing the co-relationships between the dimensions of the vector, seeking to identify semantically related terms and, using a mathematical technique called rotation, seeking to bring similar dimensions closer together. In some cases, synonyms or semantically similar terms are placed in the same dimension, thus reducing problems related to differences in vocabulary.

Another method is natural language selection, where syntactic and semantic techniques are used to analyse documents in search of the most relevant terms. According to Guthrie (1996), by using a well-defined grammar for a specific domain, it is possible to carry out a syntactic analysis of not very complex sentences, since the elements that make up the sentences usually have defined syntactic positions. By assigning weights to the terms for each position, it is possible to make them more or less relevant. Another way is to select terms according to their syntactic category (subject and complements). However, this technique requires a knowledge base containing all the syntactic categories in a grammar.

The actual text mining stage aims to extract patterns by applying a method. The main tasks involved in the text mining process found in the literature include:

- **categorisation:** text categorisation consists of analysing the content of a document in order to identify categories that can classify documents based on the context of the information they contain (RIJSBERGEN, 1979). Artificial intelligence techniques such as machine learning, decision trees and Bayesian classifiers are used in this text mining task.

- **summarisation: a** mining task that identifies the most relevant terms from a set of documents and creates a summary aimed at reducing (summarising) a document while maintaining the meaning of the text (FAYYAD et. al., 2002).

- **clustering:** this text mining task seeks to bring together a set of documents according to common characteristics, based on analysing their content.

Text clustering criteria take into account similarity measures of terms found in the documents. The literature emphasises that among the most widely used similarity measures is the *Euclidean* (COLE, 1998) (GUO et. al., 2002).

The final stage is analysing the results, which consists of interpreting the results obtained from the mining process. In this stage, the results obtained in the text mining process are evaluated to see if they meet the proposed objectives. In order for the results to be evaluated, visualisation techniques are used, supported by tools and experts in the field of knowledge, with the aim of validating the results obtained. Fayyad et. al. (2002) describe that the activity of analysing and validating data is based on statistical foundations, which provide an overview of the state of the data at a given time, through the application of concepts such as standard deviation, variance and mode. According to the author, the set of appropriate and most widely used techniques for analysing data models is based on graphs, tables and charts. Grinstein and Ward (2002) point out that visualisation systems enable data classification and the use of visualisation techniques for visual perception and user interaction with the data set extracted from the mining process.

Having explained the concepts involved in text mining, the stages and tasks involved in the process of extracting interesting patterns and new knowledge from a set of textual data, we will now present an overview of the use of data mining in educational applications. This will be seen in the next section.

5.2 Text mining in educational applications

Currently, the main innovation factor in traditional teaching systems is the introduction of new technologies (Ha, 2000). In this sense, education with the support of web-based systems is a technology that plays an innovative role in relation to traditional teaching models, since it is a form of computer-assisted instruction that is independent of the specific location of both the teacher and the student. It is also independent of hardware and operating systems (Brusilovsky, 2003).

The use of the Web as a mechanism for communication and collaboration has gained considerable importance and many courses have been implemented in recent years. However, although the web environment is convenient and offers facilities for teaching and learning, most of the courses available use static learning materials, which do not take into account the individuality and diversity of students (Romero, 2007). Each student has a learning profile which often does not match the tools used by the teacher. Static material offers a watertight resource, often not allowing the student a learning strategy suited to their individual profile. As a result, there is a need for mechanisms that make it possible to individualise learning, through solutions that adapt to the needs of each student.

This is how adaptive and intelligent web-based teaching systems were conceived. These systems are presented as a possible solution to an attempt to personalise learning, taking into account the objectives, preferences and prior knowledge of each student. In these systems, alternatives are presented to the learning modality based on static material, where the student must use only what is made available in the course's virtual space. Adaptive and intelligent systems seek to interact with the student in an attempt to adapt to the individual needs of each person. These systems are the result of the evolution of intelligent tutoring systems and adaptive hypermedia systems (Brusilovsky, 2003).

According to Romero (2007), although adaptive and intelligent systems have resources to make learning more flexible and adaptable to students' individual needs, data mining techniques can make *web-based* learning environments more effective. Such effectiveness is due to the fact that data mining techniques in educational systems can help educators discover useful information for making decisions and directing their pedagogical strategies, as well as helping them control and refine their teaching approaches.The application of data mining techniques in educational systems follows a cycle of hypothesising, testing and refining phenomena observed during the course of a course supported by a virtual learning environment. Figure 3 illustrates the cycle of activities involved in applying data mining to educational systems.

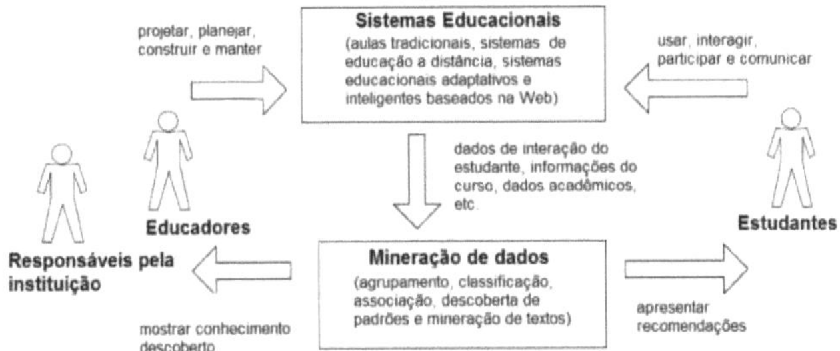

Figure 3 - Data mining in educational systems

Source: Romero (2007)

The figure above shows the interests of each group of individuals with regard to the data generated by educational systems. Note that managers and educators, each with a specific objective, are interested in discovering knowledge that provides support for planning and maintaining educational systems. In addition, knowledge generated from stored data can be useful for providing recommendations and guidance to students, who use the system for interaction and communication. Educational systems keep data about courses, teachers, students and interactions that can be useful for improving learning processes in virtual environments. In this way, educational data mining (EDM) can be targeted at the different elements that use the system, according to their particular interests.

- Student-orientated EDM: in this type of application, data mining is used to recommend activities, resources and good practices with the aim of enhancing learning. This can be done by suggesting shortcuts or links to follow, based on the learning experiences of other students, or even on tasks produced by other similar students.

- Teacher-orientated EDM: this application is geared towards obtaining greater *feedback on* instructions, evaluating the structure and content of courses, as well as the effectiveness of the learning process. Another focus of interest for this application is classifying students into groups based on monitoring and guidance needs, finding regular and irregular learning patterns, finding frequent errors, finding activities that are most effective, finding information for adapting and customising courses.

- EDM oriented towards educational managers: the application of data mining in this case is related to the production of knowledge that will serve as a basis for improving and adapting institutional websites, with the aim of making them efficient and in line with user behaviour. In this application, the emphasis is on improving institutional resources, such as human and material resources, leading to improved educational programmes.

Both traditional face-to-face teaching systems and web-based distance learning systems have data sets generated from interactions between the individuals, resources and technologies that make them up. The datasets generated by these systems can harbour useful and relevant information for educators and administrators of educational institutions, as well as for the student. Therefore, applying data mining techniques to educational systems can bring benefits for improving various aspects of these systems. The application of data mining techniques in these systems must be treated specifically for each type of system. This is necessary due to the fact that each model has specific data sources and objectives.

According to Romero (2007), educational data mining can be applied to data from two types of systems: traditional classroom-based systems and distance learning systems. This classification should be made since both have different objectives and data sources. As such, each type of system requires appropriate mining techniques.

Traditional face-to-face teaching is the environment most widely adopted by education systems. In this modality, exchanges between students and teachers take place face to face. In conventional classrooms, teachers carry out the instructional process by monitoring student learning and seeking to reinforce the assimilation of content by observing, recording and analysing student performance during the execution of tasks. Teachers can also use information about the course, such as: objectives, profile, curriculum, as well as information about student care. To do this, the teacher has at their disposal various sources of information that an educational institution produces, such as: traditional databases containing information about the student's career, school calendar, class constitution, academic data, etc.

Data mining techniques can help every element of the education system. Educational managers can use the knowledge generated by the system to forecast enrolments for class composition and timetable formatting, as well as forecasting enrolments and requirements for admission to courses. Students can obtain information about the course they want to study in order to better help them make their choice and predict their performance. For teachers, knowledge generated by the system can be useful for evaluating experiences that can result in more effective overall learning, checking and comparing the performance of different classes in the same course, analysing groups of students, etc.

Some works on educational data mining, focusing on traditional systems, can be found in the literature, one of which was developed by

Ma et. al. (2000) where the concern is to analyse students by identifying those who have poor potential for a particular course. The results of the analysis are used to suggest reallocating the student to another course of interest. To do this, the authors used a student scoring model based on association rules.

Another work, developed by Luan (2002), uses supervised and unsupervised clustering and prediction algorithms with the aim of guiding educational institutions in the allocation of resources and personnel, through proactive management in the pursuit of student development.

Distance education is a type of education system where methods are used to access the educational modality where the student is distant from the teacher, both spatially and temporally. A special feature of distance learning systems is the lack of face-to-face contact between teacher and student. There are many forms of distance learning systems, however, one of the most widely used is the provision of a virtual learning environment based on the *Web,* through virtual teaching-learning environments (VLEA).

An AVEA is a platform that provides a variety of channels and tools to facilitate communication and information sharing between participants in a course. Such systems allow educators to distribute resources and information to students, as well as producing material, preparing assessments, discussion forums, chat rooms, file storage areas, messaging services, etc.

The AVEA stores a large volume of data resulting from student activities over time. Among the activities carried out by students are: reading and writing material, carrying out assessments, taking part in debates and reflections and exchanging information between peers. To store information on student activities, the systems use a database, where data on student information and profile, results obtained in assessments, student interactions, among others, are stored. As such, data mining can result in the discovery of useful patterns for evaluating activities, as well as for teachers to analyse and learn how students' learning behaviour is in such environments, as we can see from the following works.

Lei et. al. (2003) propose the use of exploratory functions for data mining on the Web, focusing on a better understanding of the student in order to develop a formative assessment process, with the aim of improving Web-based instructional projects. With this in mind, the authors suggest analysing sequential patterns related to student learning activities, seeking important *feedback* for the teacher.

Talavera and Gaudioso (2004) investigated the use of clustering techniques to identify patterns that reflect student behaviour in data from virtual course management systems. The work focussed on the study and definition of data model patterns to represent interactions in the course environment.

The work by Dringus and Ellis (2005) analysed discussion forums using mining techniques. In their work, the authors emphasise the use of indicators of student participation in the discussions promoted through the forum, with these indicators being used to monitor student performance. The authors state that mining techniques can be used to help teachers analyse the progress of student activities.

Ventura et. al. (2008) present a study on the development of a *framework* to help teachers analyse the measures generated by assessment rules in educational data mining, supporting the creation of new rules by the teacher.

The combination of intelligent agents with data mining techniques using Bayesian networks is proposed in the work by Schiaffino et. al. (2008). The aim of this work is to observe the behaviour of students in *online* courses in order to automatically create a student profile. This profile is obtained by automatically detecting the student's actions in the environment. With this, the system will suggest personalised actions in the course for the student, with the aim of developing the student during their learning activities.

Garcia et. al. (2009) describes an iterative methodology for developing and maintaining web-based courses, the aim of which is to increase the effectiveness of the course. To this end, the authors emphasise the use of association rules to discover information of interest from data resulting from student interactions. A collaborative system is used to share the scores of the recommendation rules obtained by teachers with similar profiles, as well as by other education professionals.

Azevedo et. al. (2011) carried out a qualitative analysis of messages posted by students on discussion forums. The authors applied text mining techniques based on graphs to identify the degree of relevance of the messages posted by students. According to the authors, in this way it is possible to identify students who need more help.After an overview of the use of mining techniques in educational applications. We begin by contextualising this research with the process of text mining. In the next section, we will describe how text mining techniques are used in the scope of this thesis in order to achieve the proposed objectives.

5.3 Text mining in the context of this research

In the context of this research, we are using text mining technology to find evidence of mediation in virtual teaching and learning environments. To this end, we propose the use of computer models to detect mediation categories arising from textual interactions carried out by subjects using *online* environment tools.

Therefore, in this research we used a text mining methodology to classify the textual interactions of participants in a course, in order to identify the mediation categories of each interaction. To this end, we combined supervised learning technologies to develop an inference model. The aim is to design a model that correctly classifies known data and new examples based on the knowledge acquired. According to Feldman and Sanger (2007), classification is one of the most common data analysis processes in text mining. Its aim is to create a model based on the mapping between previously known classes and a set of textual data. This model is used to automatically determine the classes in new sets of unknown textual data.

Witten and Frank (2005) explain that among the Artificial Intelligence methods and techniques applicable to text mining, machine learning can be employed. This technique consists of systems capable of performing automatic knowledge acquisition. Therefore, in this thesis we used machine learning techniques to build classification models.

The work by Raminelli (2009), also part of the MEDIATEC project, carried out a study of classification algorithms to determine which would be used in the research involved in the project. As such, the work focussed on probabilistic classification methods. The work showed the good performance of Bayesian inference algorithms, which were used in this research.

In this way, a system was developed to map the signs of mediation and apply it to data from textual interactions in an *online* course held in the Moodle environment. To this end, an extension was developed for Moodle, which, coupled with the environment, allows the mediator to monitor the process of developing pedagogical mediation during the course.

The mediation evidence mapping system applies Bayesian inference models to the data resulting from the participants' interactions, classifying the categories of mediation carried out during the course. Generating information in the form of graphs and reports on the mediation process developed over the period.

The next chapter describes the methodology used in the research work, detailing the case study applied in this research to validate the results obtained by using the system for mapping signs of mediation in virtual teaching-learning environments.

CHAPTER 6

RESEARCH METHODOLOGY

This thesis involved applied technological research, the aim of which was to develop a system for mapping the levels of mediation carried out by teachers and tutors in a virtual teaching-learning environment.

Knowledge was also produced about the application of this research to pedagogical mediation in virtual teaching and learning environments. In this sense, we aim to apply the knowledge that underpins theories on pedagogical mediation to the design of a technology for virtual teaching-learning environments.

The knowledge generated is in the form of documentation on the process of designing the eMediation system proposed in this thesis:

- **Requirements analysis documentation:** through the use case diagram and descriptive tables of the eMediation system's requirements gathering process.

- **Documentation of conceptual and domain modelling:** drawn up using the class diagram, where we delineate the functions of the eMediation system and its interfaces with other subsystems.

- **System design documentation:** all documentation relating to the design of the application, involving the functional, navigational and structural aspects of the eMediation system.

- **Deployment documentation:** involves the deployment aspects of the eMediation system in the Moodle environment. The resources needed for the application to work are also described.

We also emphasise the documentation of the analysis of the results obtained during the application of the case study.

In addition to the technology developed, we also propose to disseminate new knowledge inherent in the process of developing this technology, as well as the new knowledge associated with the mapping model of mediations carried out in the Moodle virtual teaching-learning environment by the eMediation system.

Jung (2004) explains that determining the type of research follows the following methodological sequence:

- **first step:** understand the nature of the research, i.e. whether it is basic or applied (technological) research;

- **The second step is to** determine the research objective. In this case, the research can be exploratory, descriptive or explanatory;

- **third step:** choosing the procedure for carrying out the research.

In terms of nature, this research is characterised as applied, as it aimed to design a technology, as well as apply and evaluate the results of applying this technology through a case study. In this sense, the result of the research is a detailed description and analysis of the process of mapping indications of student learning mediation raised by the eMediation system, during the empirical exploration of its functionalities in the Moodle virtual teaching-learning environment.

In terms of the research objectives, this was a qualitative study with an exploratory technological basis, as we sought to investigate the phenomenon of pedagogical mediation in virtual teaching-learning environments. As for the methodological procedures, the research involved technological development to examine the learning environment with a view to understanding pedagogical mediation. Thus, empiricism was the essence of this research, through the use and combination of technological components with theoretical educational conceptions we sought to achieve the results of the investigation. The set of actions for comparing, including and integrating technological components led to the process of materialising the idea for solving the research problem.

The use of a real case study to apply empirical research of a qualitative and exploratory nature was justified in this research, with a view to analysing the pedagogical impacts of implementing technological innovations in virtual teaching-learning environments.

The next section describes the case study applied during the research work described in this doctoral thesis.

6.1 The case study

According to Gil (1999), a case study is based on an in-depth and exhaustive analysis of a few elements, so that it is possible to gain a broad and detailed knowledge of a fact, which would not be possible with other research designs. It is an empirical study that examines a phenomenon within its context of reality. Yin (1994) adds that a case study is appropriate when guiding questions such as "why", "how" and "in what way" are involved in defining the research problem.

The thesis sought to answer questions such as "how" and "in what way" (how can the mediation process be perceived and managed in the different interaction spaces of the virtual teaching-learning environment), (how can textual data mining be used to map interactions and survey the levels of learning mediation in virtual teaching-learning environments?

The case study was applied to a course on media in education at the Interdisciplinary Centre for New Technologies in Education (CINTED) at the Federal University of Rio Grande do Sul. From the first of August two thousand and eleven to the thirty-first of October two thousand and eleven. During the development of the subject Pedagogical Practice in an ICT-supported environment.

The research was carried out in a virtual teaching-learning environment, since we are looking to specify, design and develop a system and, to do this, we need to control aspects that will interfere in the development of the research, such as: methods, technologies, tools and software components that will be used to specify the architecture and develop an application to explore the phenomenon of mediation through the application of a case study.

Therefore, we need to use a virtual teaching and learning environment for the study. In this case, we used the Moodle environment[7] . This environment was chosen because it is widely used in subjects in the postgraduate programme in Informatics in Education at the Federal University of Rio Grande do Sul to support teachers' activities. Another determining factor for adopting the Moodle environment in the exploratory studies carried out in the research is its popularisation and constant evolution. It is also open source software, which allows it to be freely studied and its technological resources expanded. This last factor was decisive, as the system developed in this research had to extract data from the Moodle database, as well as allowing direct interaction between teachers and tutors. To this end, it was developed as an extension to the Moodle environment.

Not all the tools in the Moodle virtual learning environment were explored in the case study, although the system can contemplate pedagogical mediation across the full range of available tools. The eMediation system encompasses the mediation space of the Moodle environment as a whole, based on interactions between participants through the various tools: forum, chat and Wiki. These tools were chosen because they allow textual interaction between participants. In the context of the course applied in the case study, we will detail the pedagogical strategies adopted by the teacher of the course during the use of the tools by the students, as well as the tasks proposed for their use in the course.

A pilot study was carried out before the system was applied in the case study. This pilot study was important for verifying the process of capturing signs of mediation by identifying the levels of mediation of the learning of the students involved, as pointed out in the works of (PASSERINO et. al., 2008; 2000), (KOCH, 2009) and (DIAZ, 1993). The pilot study was *Ex-Post-Facto,* as it used data from a course that had already taken place. In addition, we wanted to check the prototype's behaviour during its integration with the other software components used to implement the eMediation system.

---

[7] Moodle is an online course management system that was developed to encourage interaction between students in order to increase learning through the construction of ideas based on experiences arising from social interactions and elements of the virtual environment (DOUGIAMAS and TAYLOR, 2003).

6.2 Data collection technique

Yin (1994) highlights six important sources for data collection in case study-based research:

- documentation;

- records in archives;

- interviews;

- direct observation;

- participatory observation;

- and physical artefacts.

The author emphasises that the sources are complementary and combining them makes for a good case study. In addition, the procedures for gathering evidence must be managed independently, ensuring that each source can be used appropriately by the researcher.

This research used direct observations of the processes involved in implementing the eMediation system in the Moodle virtual teaching and learning environment. As well as its behaviour during the capture of signs of mediation. Using observation, we sought to identify evidence of mediation in the environment by identifying the level of pedagogical intervention needed by students based on the categories raised in the works of (PASSERINO et. al., 2008; 2000), (KOCH et. al., 2009) and (DIAZ, 1993), using text mining algorithms.

We will also use a semi-structured interview with the subject tutor. We will use the semi-structured interview technique because we are looking for more comprehensive answers to our questions. By using the interview, we are looking for elements that could be useful in analysing the pedagogical impact of applying the mediation evidence mining system in the *online* environment. Another important aspect of the semi-structured interview is that it is flexible in terms of duration, allowing for an in-depth approach to the topic being investigated. The interaction between the interviewer and the interviewee in a spontaneous way allows for the exploration of subjective aspects about the students' performance throughout the course.

To validate the thesis, we triangulated a semi-structured interview with the course tutor; direct observations by the researcher of the mediations carried out in the environment during the course; and the information generated by the system on the mediations carried out in the environment.

To conduct the research, we organised the work into a series of stages, as described in the next section.

6.3 Research stages

The first stage of this research involved a literature review and study of aspects of mediation from a socio-historical perspective. We also studied distance education with the support of *online* educational environments, more specifically with the use of virtual teaching-learning environments, highlighting their potential, problems and future prospects. Finally, we carried out a study on text mining computer technology and its applications in education. With these preliminary studies, we wrote the literature review and contextualised it with the research developed in this doctoral thesis.

Based on the theoretical framework, we moved on to the second stage of the research, which was the study of text mining technologies and their application in surveying evidence of mediation in a virtual teaching and learning environment. During the study we carried out research into computer tools to support the mining process, as well as libraries of software components for building applications. After the study, we applied Bayesian inference algorithms using a text mining tool. The aim was to obtain an inference model to be applied to texts arising from interactions between students and mediators in the virtual environment.

The third stage consisted of gathering requirements for the mediation mapping system. This stage was concerned with defining the documents resulting from the requirements analysis process, which served as the basis for the system's design. During the requirements analysis stage, a conceptual and domain model of the system was drawn up. Next, system design documentation was drawn up, describing the system's functions, architectural aspects and its implementation.

The fourth stage involved building the system by coding the functions identified in the requirements gathering and design phase. Tests were also carried out on the application using a pilot study to check the software's performance in the process of capturing evidence of mediation. This stage also involved a study of the logical structure of Moodle's architecture. With this, we sought to understand how the software was conceived, exploring the components involved in its architecture. In addition, we also needed to research the coding standard for extensions to the environment, since we needed to develop a system that worked as an extension to Moodle. Another important aspect was the study of integration between technologies, as the technology used to develop the system is different from the technology used to build the Moodle VLE.

The fifth stage of the research involved implementing the clue mapping system in the Moodle environment. Once the system was in place, we applied it to the *online* course which is the subject of the case study proposed in this research. The system was applied during the course Pedagogical Practices in ICT-Supported Environments. In three data collection periods, according to the structure defined in the course planning.

The sixth stage of the research analyses the data and results obtained from the case study. To this end, three data analyses are carried out. The first analyses the data collected during the application of the e-mediation system. The second analysis is carried out by observing the data resulting from the participants' interactions from Moodle's own database. The third analysis *is carried out from* the point of view of the course tutor. The three analyses were triangulated in order to validate the results of the research. Finally, we moved on to the stage of analysing the results obtained by applying the system for mapping signs of mediation in VLEs. The results are presented in the next chapter.

CHAPTER 7

RESEARCH AND RESULTS

In this chapter we will describe in detail the stage of applying the system in the case study and analysing the results. This phase was the focus of the research work, whose application of the system's functionalities was aimed at mapping interactions and identifying the levels of learning mediation throughout the development of the course. This stage constitutes the analysis of the results of the research work.

7.1 Context of the subject under investigation

In this section we present a detailed description of the structure of the course that is the subject of this research case study. This description is necessary so that we can get to know and understand the context in which the research was carried out.

Initially, we needed to define a period for applying the system to extract information about the pedagogical mediation carried out in the course. As the Media in Education course is structured around several subjects, we had to define the period of the course relating to the subject **Pedagogical Practice in an ICT-supported environment.** Since this subject was adopted for the case study. Therefore, the period of application was the same as the duration of this subject, which took place between August and November of two thousand and eleven.

The subject Pedagogical practice in an ICT-supported environment was structured in three stages, or parts:

- **Part I -** Environments, pedagogical practices and technologies;

- **Part II -** Authorship, interactivity and collaboration;

- **Part III -** Personal learning communities and environments.

Below is a detailed description of the teaching resources and assessment tools used during each part of the course.

7.1.1 Part I: environments, pedagogical practices and technologies

The first part of the course lasted approximately four weeks, starting in August two thousand and eleven. The aim of this first part was to analyse the teaching-learning process. Videos, texts and reviews on the teaching-learning process were used as teaching resources.

In the first week, students were presented with a text on teaching and learning and a book review. Immediately after reading the text, the student had to enter the forum "Teaching and learning process: today at school". To assess this first part, some reflections were made on questions posted on the forum tool. Students had to make a significant contribution to the forum, either by starting a debate or by contributing to a colleague's position. The following questions were posted for reflection in the forum:

- does teaching imply learning?

- And the other way round, does learning imply teaching?

- What elements are present in a learning environment? Give an example.

The students should discuss with their colleagues, agreeing or disagreeing, but presenting elements to support their discussion. These elements could be derived from the student's teaching practice or supported by authors present in the theoretical framework.

In the second week the students had to watch the video: "Learning to learn". Then they returned to the forum "Teaching and learning process: today at school" and answered the following questions:

- does teaching imply learning?

- And does learning imply teaching?

- What elements are present in a learning environment? Give an example.

In the third and fourth weeks of this first part of the course, students had to reflect on the use of technology in the teaching and learning process. To do this, some videos and texts were used, as well as the presentation of concepts, paradigms, visions and realities that affect public policies and experiences of using technologies in schools.

At the end of the fourth week, the students had to re-enter the forum and post more contributions on the topic they had worked on in the first part of the course.

7.1.2 Part II: authorship, interactivity and collaboration

The second part of the course lasted approximately eight weeks, from September to October two thousand and eleven, focusing on learning with technology, as well as the relationship between the themes of authorship and collaboration. The following objectives were set:

- understanding of concepts such as interaction, interactivity, authorship and collaboration with the support of technology;
- knowledge of some technologies that favour forms of collaboration and authorship;
- analysing the potential of technologies for configuring new educational processes with the support of technologies.

The activities for this stage were scheduled to take place weekly over the eight weeks planned. For this part of the course, in addition to the forum, Moodle's Wiki tool was also used as an assessment tool. Texts, guides and tutorials on the tools used were used as teaching resources for this part of the course.

The fifth week of the course began with a reflection on two teaching-learning situations with computer support in the classroom. This reflection should be based on previous discussions about the teaching-learning process covered in the first part of the course. Students should then analyse the two situations presented on video, identifying the following aspects in each one:

- the concept of education that underpins each practice;
- role of the teacher;
- role of the student;

- the role of technology in the educational process in the two videos.

All of the students' questions during the activity should be posted in the forum: "Authorship and collaboration with technologies".

In the sixth week, in the second part of the course, students were asked to discuss the concepts of authorship and collaboration using technology. This discussion should be based on the following questions: is there authorship and collaboration in the situations presented in the two videos? If so, should the student describe what kind of authorship and by whom? If not, why?

As this week was entirely based on discussions via the forum on the issues raised, each student had to make at least two relevant contributions to the forum.

In the seventh week of the course, two texts were used to support the resumption of discussions on the elements needed for a technology to enable authorship and collaboration.

Weeks eight and nine were used for the students to experiment with the use of authoring and collaboration tools. To do this, the students were introduced to Moodle's Wiki tool. As a task for learning the Wiki, group work was proposed. The students had to form groups of three.

Each group had to collaboratively produce a text summarising what had been covered in weeks five to seven. This text should include the concepts of authorship and collaboration, using diagrams, videos, pictures or concept maps. An important point to emphasise is that the work should be done collaboratively and not simply by subdividing the group. Collaboration should be produced by adding, altering, correcting or deleting elements from the document.

Questions should be directed to the general class forum. A number of tutorials were made available to support the use of Moodle's Wiki tool. From the tenth to the twelfth week of the course, activities were proposed to develop an educational practice with the support of authoring and collaboration technologies. To this end, each group of students was asked to devise a lesson plan that would enable the use of some kind of authoring and collaboration tool. To facilitate the group's work, some tools were suggested, in addition to the Moodle wiki. The choice of authoring and collaboration tool was at the group's discretion. What was important was that the lesson plan emphasised the promotion of authoring and collaboration.

7.1.3 Part III: communities and personal learning environments

After addressing pedagogical practice with the support of technologies and the issue of authorship and collaboration as a learning resource, the third and final part of the course was presented. At this point, a recent phenomenon in today's society, which is also present in education, was discussed: social networks on the Internet. This part also dealt with a new concept that has recently emerged: personal learning environments (PLE). Therefore, the objectives set for this third stage of the course were:

- discuss the potential of social networks for educational use;

- to understand how ELPs are constituted and their possible uses in education.

In the thirteenth week of the course, tasks were proposed to explore and get to know the topic of networks, communities and personal learning environments. To this end, two texts were made available for reading on the topics proposed in this part of the course. As a task, a collaborative writing of a glossary of terms on the topics covered was proposed to help in understanding the theme addressed at this stage.

In the fourteenth week, a synchronous session was held using Moodle's chat tool. In this session, the concepts of social networks were presented and discussed. The discussion that had begun in the synchronous meeting was continued in the forum: "Social networks and education - limitations and potential".

In the last week of the course, an initial reading was proposed on a topic that is emerging in the international scientific community: personal learning environments. To this end, two reading references were presented to contextualise the students with the topic. In addition to the texts, an explanatory video on the subject was also shown.

As a task, the students were asked to take an individual stance on the issue. To do this, each student had to post contributions in the forum "Social networks and education - limitations and potential".

After learning about the structure of the course that was the subject of the case study involved in this research, we moved on to analysing the results obtained from the case study. This will be presented in the next section.

7.2 Analysing the data

Recalling our research question: how can text mining technology and socio-historical epistemology provide elements to support pedagogical mediation in virtual teaching-learning environments? To this end, we adopted a strategy for analysing the data resulting from the textual interactions of the course participants.

When analysing the data, we took into account the period of the course when the tasks were applied, which required the participants to interact more with the tools in the Moodle environment. With this in mind, we could guarantee a greater amount of data for analysis when applying the mapping algorithm. In this way, we delimited some spaces for applying the system within the period of the course. This decision was due to the fact that we also wanted to check the evolution of the student learning mediation process during the course. We therefore defined three periods for mapping data and applying the inference mechanism to survey the levels of student learning mediation. The periods defined were: August, September and October. The period during which the course was taken lasted until the end of November. However, the period selected was sufficient because most of the activities that would feed the Moodle database with the textual interactions needed for the eMediation system were carried out during this period.

In addition to the periods of application of the system, we also needed to select a specific class, since the Media in Education course is applied in several centres, where each centre participates with one, two or three classes. We therefore had to decide which centre and class to adopt for the case study. This decision had to take into account the most participative classes, with the highest number of contributions in the environment, i.e. those that interacted more with Moodle. This aspect was decisive for the choice of class, as we depended on it for the research to be successful. In this sense, the choice was supported by a previous analysis of the classes at each pole based on interactions carried out in periods prior to the subject of the case study. We then analysed the interactions of four classes based on data collection, so that we could choose a more interactive class. We therefore decided to analyse class T2 at the Porto Alegre pole.

This analysis aims to answer the guiding questions of this research: 1) How can the process of mediation be perceived and managed in the different interaction spaces of the virtual teaching-learning environment? 2) How can text mining be used to map interactions and identify signs of mediation in virtual teaching-learning environments? 3) How can information resulting from mapping be made available to help the teacher with the task of pedagogical mediation in a virtual teaching-learning environment?

Therefore, in order to answer the guiding questions and achieve the objectives of this research, we will begin to describe the results from the following points of view:

- An analysis of the results of the application of the eMediation system and the graphs and reports generated by the tool.

- An analysis of the interactions carried out in the environment by the researcher himself.

- An analysis of the tutor from an interview.

A triangulation of the results of these analyses will allow us to validate the results obtained by the system during the proposed case study, leading us to the conclusion of the research being carried out.

Another relevant aspect of this research is the delimitation of the subjects to be analysed in the case study. The class adopted for the research consisted of thirty students. So we came to another decision point: to analyse the mediation of the learning of which students? To do this, we established the following criteria for selecting the students to be analysed.

- Number of interactions in the different tasks proposed throughout the course in the forum, chat and Wiki tools.

- Number of accesses to the environment, permanence and resources accessed by the course participant over the observation period.

- Diversity of mediation levels indicated by the eMediation system during the mapping stage of the participants' interactions. So that we could monitor whether or not there had been an evolution in the mediation of student learning.

The criteria for choosing the students were supported by the information generated by the mediation graph based on the interactions of the course participants.

This graph is useful for identifying the students who carried out the most interactions. We can also see from the graph the diversity of levels of mediation of the interactions carried out by the student. We can make a comparison between the students to establish which are the most participative, at what level of mediation their interactions are and the diversity of the participants' mediations. We would like to emphasise that the mediation graph will be further detailed in the section on analysing the application of the e-mediation system within the scope of this thesis.

Another important aspect to note is that we extracted mediation graphs over three observation periods in August, September and October. This allowed us to monitor the evolution of the levels of mediation in the participants' interactions over the course of the course.

In this way, we selected some students with the highest number of interactions and others with the fewest. In addition to the number of interactions, it would also be interesting to analyse those students who show varying levels of interaction mediation. But we couldn't leave out those students who showed a linear level of learning mediation, with few variations. To this end, we sought to see how learning mediation evolved over the periods observed. To do this, we selected four subjects to analyse over the course of the subject, which in this thesis we will refer to as subject **A, subject B, subject C and subject D.**

In the following sections we will begin to analyse the results from the three perspectives listed above. Starting with a preliminary analysis of the subjects analysed in the research.

7.2.1 . Profile of the analysed subjects

The purpose of this section is to give the reader a profile of each subject observed during the course of the research. This profile presents aspects such as the academic background and experience of each subject. In addition, in selecting the subjects, we took into account the previously established criteria.

- **Subject A** works as a teacher in the state of Rio Grande do Sul, training professionals in the field of education. Subject c also has experience in coordinating computer labs, research and learning using the Internet.

- **Subject B** is a teacher who works with children in kindergarten A and B in one school, and with children in the first grades of primary school in another. She also has an administrative role in a school, holding the position of headmaster. Both schools are in the interior of the state of Rio Grande do Sul.

- **Subject C** is a teacher from the state of Rio Grande do Sul, working in two schools in the capital in primary and secondary education. As well as taking a postgraduate course in Media in Education, she also takes part in extension courses aimed at developing projects for schools.

- **Subject D** is a secondary school teacher in the Rio Grande do Sul state school system and in a school in the interior of the state. In addition to the Media in Education course, he has other academic qualifications and professional certifications.

After describing the general profile of the subjects, we went on to analyse the results of mapping their levels of learning mediation during the course of Pedagogical Practice in an ICT-supported environment.

The results of the analysis of the research carried out in this thesis will be described from the three points of view listed above. We begin by analysing the results indicated by the system in the next section.

7.2.2 Application of the e-Mediation system

The results of the application of the eMediation system were analysed on the basis of the information generated by the reports and graphs during the three mapping periods. We began by analysing the results presented by the graph of the participants' level of mediation. This is a general graph in which the series represent the interactions carried out by the course participants in each inferred mediation category. The graph does not separate the interactions by tool, but generates an overview of all the textual interactions carried out in the environment.

From the mediation graph based on the participants' interactions, we have a position on the students' levels of learning mediation, based on their textual interactions in the environment. We also have an overview of the levels of mediation based on the interventions made by the mediator.

In this study we extracted information in three periods during the course, so we drew up three mediation graphs of the participants' interactions. The first period of information extraction took place from the first of August until the last day of the same month. During this period, the first part of the course took place: "environments, pedagogical practices and technologies". And as an activity of the course, discussions were proposed in the forum: "teaching and learning process: today at school". Figure 4 illustrates the graph extracted during this period.

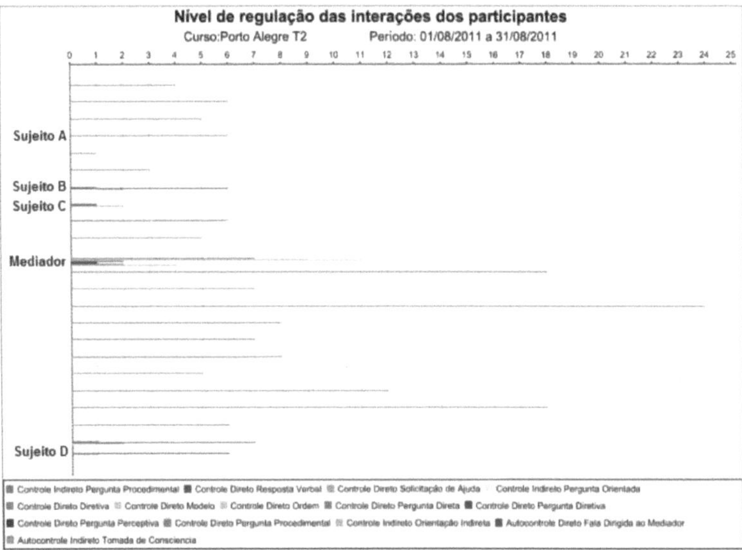

Figure 4 - Mediation graph of interactions in August/2011 Source: author's own work

Two hundred and thirty-eight interactions took place between all the participants during the period. Thirty-seven pedagogical interventions were made by the mediator. Among the interventions made by the mediator, direct and indirect control interventions were inferred.

Among the direct control interventions mapped by the system were: direct control in the form of directives (7), direct control in the form of a model (9), direct control in the form of orders (11), direct control in the form of a direct question (2), direct control in the form of a directive question (1), direct control in the form of a perceptual question (1), direct control in the form of a procedural question (2). As for indirect control interventions, the following were mapped: indirect control in the form of indirect guidance (4). This first analysis shows the use of direct control in the form of orders from the mediator.

While analysing the interactions of the subjects of the research proposed in this thesis, we found the following patterns in their interactions:

- **Subject A:** the student maintained a pattern in the level of mediation of her interactions in the environment during the first period of data mapping.

The student had six interactions classified in the indirect control category in the form of a procedural question.

- **Subject B:** the student showed a variable pattern of mediation levels in her interactions during the period analysed. Among her interactions we identified: one interaction of the indirect control type in the form of a procedural question, six interactions in the direct control category in the form of a verbal response, two direct control interactions in the form of a request for help and one of the indirect control type in the form of a guided question. It can be seen that the student resorted more to responses to direct controls carried out by the mediator.

- **Subject C:** the student also showed a variable pattern in the level of mediation of her interactions during the period. Of these interactions, there was one indirect control interaction in the form of a procedural question, one direct control interaction in the form of a verbal response and two direct control interactions in the form of a request for help. The student posted the fewest contributions in the period, only four posts on the forum. It can be seen that the student was concerned about help in this first instance.

- **Subject D:** the student showed a more constant pattern, like subject A, with only one interaction in the category of indirect self-control in the form of awareness and six interactions of the direct self-control type in the form of speech addressed to the mediator. We noticed that the student turned to the mediator in anticipation of actions to be taken, showing self-control of his actions.

  In a preliminary analysis, we can see that **subjects B** and **C** are at the direct control level in the process of mediating their learning, since most of their interactions were mapped to this category. We also noticed that **subject A had** the most linear pattern of the group, with all his interactions remaining at the indirect control level, using procedural questions addressed to the mediator. Subject **D,** on the other hand, is in the self-control category in this first mapping period, since all of his interactions were mapped to this category.

We now move on to an analysis of the results of the mapping of interactions in the second period of the course, which lasted the whole of September and part of October. This was the second part of the course: "authorship, interactivity and collaboration". Activities for this stage of the course included discussions in the forum "Authorship and collaboration with technologies" and collaborative writing using Moodle's Wiki tool. Figure 5 shows the graph extracted for the period.

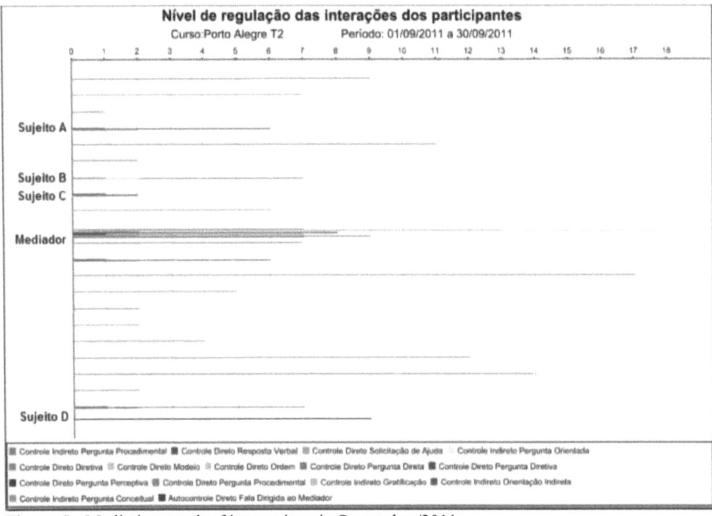

Figure 5 - Mediation graph of interactions in September/2011

Source: author

Two hundred and fifty-two interactions took place between all the participants during this period. The mediator made sixty-nine interventions during the period. In this set, direct and indirect control interventions were observed. The direct control pedagogical interventions included: directives (7), models (13), orders (18), directive questions (8), procedural questions (7), conceptual questions (7), direct questions (2) and perceptual questions (1). Indirect control interventions included: indirect guidance (9) and gratification (2). We can also see from the analysis that the mediator resorted to direct control in the form of orders, since most of his interventions were identified in this category.

81

When observing the interactions of **subjects A, B, C and D,** we saw the following mediation patterns:

- **Subject A:** unlike the first mapping period, where the student maintained a pattern in the level of mediation of her interactions, in this period there were variations in the level of mediation, with a greater number of interactions at the level of direct control, such as: request for help and verbal response.

- **Subject B:** the student maintained the mediation pattern of her interactions during this period. Most of the interactions were classified as being at the level of direct control, as she resorted to requests for help and verbal responses.

- **Subject C:** the student maintained a balance in the level of mediation of her interactions. Four interactions were identified at the level of direct control and four interactions at the level of indirect control.

- **Subject D:** the student remained at the same level as in the previous period. All of his interactions were classified at the level of direct self-control, through the form of speech addressed to the mediator.

Analysing the second period of interaction mapping carried out in the course, we can see that **subject A** showed greater variation in the mediation of her interactions. During this period, the student had doubts and resorted to asking for help. **Subjects B** and **C, on** the other hand, maintained the same pattern as the previous period, where their interactions were classified at the level of direct control. **Subject D, on the other** hand, remained at the self-control level, as his interactions were speech directed at the mediator.

In the final part of the course, which took place in October, the topic of "communities and personal learning environments" was discussed. The following tasks were proposed at this stage of the course: collaborative writing using Wiki, a meeting via the *chat* tool to discuss a topic and, later, the use of the forum: "social networks and education - limitations and potential". Based on the interactions in the three tools, the texts resulting from this period were mapped. Below we present the results of the mapping of the participants' interactions. Figure 6 illustrates the interaction mediation graph.

Figure 6 - Mediation graph of interactions in October/2011 Source: author's own work

The last period of the course saw the highest number of interactions in the three periods in which the system was used. In this phase, the participants had one thousand four hundred and ninety-three interactions. Of these, the mediator contributed two hundred and thirty-two direct and indirect control pedagogical interventions. The direct control interventions were: directives (16), model (24), order (97), direct question (2), directive question (29), perceptual question (2), procedural question (7). And the following indirect control interventions: passive confirmation (1), gratification (34), directive (1), indirect guidance (11), conceptual question (7) and direct waiver (1).

When observing the interactions of the four students who were the subjects of the investigation, we noticed the following patterns of mediation levels:

• **Subject A:** the student interacted the least with the environment in this phase

of the course, there were only two interactions, where one occurrence of self-regulation was identified and another classified in the category of indirect control in the form of a procedural question addressed to the mediator.

- **Subject B:** the student carried out eighty-three interactions at this stage of the course. The vast majority of these interactions were classified in the indirect control category in the form of a procedural question addressed to the mediator.

- **Subject C:** the student had fifty-six interactions in which the vast majority, like the rest of her colleagues, were classified in the indirect control category, in the form of a procedural question addressed to the mediator.

- **Subject D:** the student contributed two hundred and sixty-eight interactions which were classified in the indirect control category, in the form of a procedural question addressed to the mediator.

When we applied the system to the data from the interactions in the third part of the course, we noticed that the level of mediation of the participants' interactions was mapped to the indirect control category. At this first moment of analysis, we cannot verify the content of the students' interactions. As such, we are not yet able to carry out a more detailed analysis of the classified texts, enabling us to get an idea of the dynamics of the mediations carried out. We therefore moved on to analysing the mediation maps, since they give us an insight into the mediations carried out in the environment between the course participants.

The mediation map provides information based on interactions in the forum tool. Using the mediation map we can observe the development of the discussions posted on the forum and check the mediation levels of each interaction indicated by the system. In this way, we can obtain a more detailed view of the general information presented by the mediation graph based on the participants' interactions.

The mediation map generates the dynamics of interactions carried out by a particular student on the course. The map shows interactions based on a discussion launched by the student and the mediations carried out. As well as interactions in which the student makes contributions, even if they are not the author. The aim is to monitor the evolution of mediations directed at a particular student. So that we can monitor the levels of mediation over the course of the mediations carried out. Therefore, below we will analyse the mediation maps of each subject in this research.We will begin by analysing **Subject A's** mediation map. To do this, we present in Chart 1 the interactions that took place during the first period of the course. It should be noted that during this period the activity of discussing the topic "teaching and learning process: today at school" was carried out using the forum tool.

Map of Mediation			
Participant: Subject A		**Period: 01/08/2011 to 31/08/2011**	
Discussion: Teaching and learning process: today at school!			
Date	Participant	Type of participant	Level of mediation
19/08/2011	Subject A	Student	Self-regulation
Contents			
I decided to comment on J's comment because my ideas are in line with those she describes in this topic. So yes, the process of teaching and learning or (teaching + learning) and not teaching-learning or teaching/learning, in other words, teaching and learning are intrinsic, there is no subtraction or division during the process, the two are interdependent and should be discussed at every pedagogical meeting.			
Date	Participant	Type of participant	Level of mediation
20/08/2011	Mediator	Mediator	Direct directive control
Contents			
Hi F. I loved your reflection. We shouldn't exclude old methods in favour of new ones. I've noticed in the school where I work that there are classes that don't adapt to the progressive trend. Shouldn't the various tendencies be mixed or tend towards one or the other depending on the profile of the class? Have we learnt nothing from students who come from a totally progressive method?			
in our childhood and adolescence? Shall we discuss these issues?			
Date	Participant	Type of participant	Level of mediation
24/08/2011	Subject A	Student	Indirect control procedural question
Contents			
I would like to pose a question to my colleagues: assuming that one swallow does not make a summer, to what extent can a teacher sustain his or her tendency if it is contradicted by the system?			
Date	Participant	Type of participant	Level of mediation
24/08/2011	Subject A	Student	Direct control procedural question

Contents
However, I would like to draw your attention to the following question: on the assumption that doctors are also human, why do we get so angry when one of these professionals makes a mistake? Does this mean that if we make a mistake with 25 humans in our classroom - a meaningful teaching and learning process - we are forgiven for being flesh and blood teachers? Why is this kind of consideration made among graduates, bachelors are taught that they have to get it right, or else they have to get it right in another way!!!

Date	Participant	Type of participant	Level of mediation
24/08/2011	Mediator	Mediator	Direct control model

Contents
Hi F. Education by example is fundamental. We see it in the classroom and also in our lives with our children and, in my case, my godson. If I don't want him to repeat a certain action, I have to be the example. I can't charge them if my practice is different. Your question is very interesting. I'm still thinking about why certain professionals don't allow mistakes. I'd never thought about it.

Chart 1 - Map of mediations carried out for subject A in August/2011

Source: author

Analysing the development of the discussions held using the forum tool makes it possible to check the posts made by **subject A,** as well as the interventions directed at him by the mediator. With this, we can analyse the level of mediation indicated by the system for each interaction between the participants in the discussion. The table shows what was indicated when analysing the mediation graph based on the participants' interactions in the first period of the course. In this case, the graph showed that the majority of subject A's interactions were classified as indirect control, in the dimension of procedural questions directed at the mediator. This can be seen from the information in the table extracted from the mediation map. **Subject A's** initial message is classified as self-regulation and his other interactions as indirect control. The mediator, in turn, makes pedagogical interventions in the form of directives and models.

Chart 2 gives an overview of the discussions in the second period of mapping the interactions in the forum. During this period, two tasks were proposed to the course participants. The first task was to post contributions in the "Authorship and collaboration with technologies" forum. The second task was to construct a collective text on the topics covered in this second part of the course, using Moodle's Wiki tool.

Map of Mediation			
Participant: Subject A		**Period: 01/09/2011 to 30/09/2011**	
Discussion: Authorship and collaboration with technologies.			
Date	Participant	Type of participant	Level of mediation
17/09/2011	Subject A	Student	Direct control verbal response
Contents			
W. You've convinced me. There really is no authorship on the part of the students. Although I kept thinking about the possibilities of meaningful construction within the teacher's initial proposal, she could have combined the use of the computer with a poster production, images from the internet, the use of the computer and a concept map in Word with what the students had learnt.			
Date	Participant	Type of participant	Level of mediation
26/09/2011	Subject A	Student	Direct control verbal response
Contents			
Really F. The lack of authorship in their work leads students to always, or most of the time, use Ctrl c and Ctrl v in their work. Students don't think about what they are copying, they just check that they are complying with the request.			
of the task by the teacher. We are also responsible and we have seen the impact of this in the video by instigating our students to be autonomous and authors, to base their answers and understand what they are doing. To be critical and aware.			
Date	Participant	Type of participant	Level of mediation
26/09/2011	Subject A	Student	Indirect control procedural question
Contents			
0 teacher's work, in addition to the theory used, must also be based on the school regulations, because after analysing an IT syllabus, I saw that certain standards predate our arrival at the school. Of course, our role as educators is also to improve our colleagues' knowledge of our field. Therefore, the syllabus must not be limited to the specific content of the computer tool (software and hardware), but it must cover the content that can be achieved in the computer class, including, most importantly, critical thinking about the content used.			
Date	Participant	Type of participant	Level of mediation
19/09/2011	Mediator	Mediator	Direct directive control
Contents			
Hi F. Very good point. To the extent that the student researches a particular item and creates it automatically, he is learning the tools of the computer and, above all, building knowledge about the content studied. Cheers J.			

Chart 2 - Map of mediations carried out for subject A in September/2011

Source: author

The information contained in the table reflects the mappings of interactions presented in the mediation graph. The mediation graph based on the interactions indicated that **subject A** carried out interactions classified in the direct and indirect control categories, as we can see from the table. In the table we can see part of the student's posts in the discussion forum proposed at this stage of the course. In **subject A**'s posts, we see verbal responses to questions introduced by other colleagues or the mediator. The results of the third and final period of interaction mapping carried out by the system are shown in Table 3. During this period, an *online* session was held on concepts on the subject of social networks and then a discussion task was proposed in the forum: "Social networks and education - limitations and potential". However, the posts centred on the previous theme "Authorship and collaboration with technologies", as can be seen below.

Map of Mediation			
Participant: Subject A		**Period: 01/10/2011 to 31/10/2011**	
Discussion: Authorship and collaboration with technologies.			
Date	**Participant**	**Type of participant**	**Level of mediation**
13/10/2011	Subject A	Student	Indirect control procedural question
Contents			
That's right F. Collaboration is one of the principles of a meaningful teaching and learning process. Interactivity is the capacity for integration between collaborative media and between the collaborators themselves. Authorship is the ability of the subject to act significantly and interactively and leave their collaborative mark on a given event, or in the case of the school, to be the protagonist in their own teaching and learning process.			
Date	**Participant**	**Type of participant**	**Level of mediation**
14/10/2011	Subject A	Student	Self-regulation
Contents			
Really M. It is necessary for us educators, based on a Vygotskyan practice, to be the mediators between what is new and our students and even our colleagues. In order to reach everyone, we need to take a collaborative and encouraging stance and provide the means for those involved in the teaching and learning process to be critical authors of their history and also protagonists of our history. Enough of just remembering the students who were a nuisance, Cheers.			

Chart 3 - Map of mediations carried out for subject A in October/2011

Source: author

During this period, the student only interacted twice on the course forum. All the interactions took place in the "Authorship and collaboration with technologies" forum. Despite the small number of posts, one post was categorised as self-regulation, based on a reflection made by **subject A. Another was classified as indirect control in the form of** a procedural question. Another post was categorised as indirect control in the form of a procedural question.

We will now analyse Subject B's mediation map, based on the information presented in Chart 4. The table gives an overview of the interactions that took place during the first period of the course.

Map of Mediation			
Participant: Subject B		**Period: 01/08/2011 to 31/08/2011**	
Discussion: Teaching and learning process: today at school!			
Date	**Participant**	**Type of participant**	**Level of mediation**
17/08/2011	Subject B	Student	Direct control verbal response
Contents			
I think that the teaching and learning process needs to be discussed in schools, teachers need to rethink their pedagogical approach so that it is a process of construction, where the student is the subject of their learning. The teacher's focus should be on the construction of knowledge, seeking to develop a critical sense of autonomy so that the student has knowledge is always informed is able to solve their problems can think about society interact with it and transform it.			
Date	**Participant**	**Type of participant**	**Level of mediation**
20/08/2011	Mediator	Mediator	Direct control model
Contents			
Hi J. Excellent reflection on the material studied. I've posted some questions in this forum to spice up the discussion. Have a look. Cheers.			
Date	**Participant**	**Type of participant**	**Level of mediation**
20/08/2011	Subject B	Student	Indirect control targeted question
Contents			
I think this problem of a lack of motivation on the part of some teachers is due to the fact that they feel undervalued, that they earn too little, that they don't have the support of their families, that they work too many hours. I see that the training offered always generates dissatisfaction on the part of some, if it takes place on Saturdays they complain, if it takes place in the afternoons they complain, if it takes place during recess they complain - what to do? I believe that this is a challenge for the pedagogical team, to talk to these teachers, to see what can be improved so that they can improve their skills.			
moments of training are learning moments, that they are not there simply out of obligation.			
Date	**Participant**	**Type of participant**	**Level of mediation**
23/08/2011	Subject B	Student	Direct control verbal

			response
Contents			
I think that teaching and learning are very closely related. When you teach, you want learning to take place, but this doesn't always happen. For teaching to lead to learning, the teacher needs to use different teaching methods and techniques in order to reach everyone, and the student in turn must be interested in learning what is being taught. Therefore, the teaching and learning environment must be motivating, where students are challenged to participate and build knowledge.			
Date	**Participant**	**Type of participant**	**Level of mediation**
24/08/2011	Mediator	Mediator	Direct control model
Contents			
Hi J. Excellent analysis based on Moran. Now we have to discuss what to do to change the situation. There is a lack of will on the part of some teachers, but work overload doesn't allow teachers to qualify either. We still have a lot to learn.			

Chart 4 - Map of mediations carried out for subject B in August/2011

Source: author

Subject **B**'s mediation map, in the first period of mapping interactions in the course, gives us a scenario of the development of interactions and their respective levels of mediation. We can see on the map a set of mediations carried out during the period for the student's interactions. Among the levels of mediation carried out, there is a significant number of direct control interactions in the form of verbal responses, where **subject B** answers questions raised in the "teaching-learning process: today at school" forum. We also noticed the occurrence of indirect control in the form of a guided question. Table 5 describes Subject **B**'s mediation map during the second period of mapping interactions in the subject. We now move on to checking the results presented.

Map of Mediation			
Participant: Subject B		**Period: 01/09/2011 to 30/09/2011**	
Discussion: Authorship and collaboration with technologies.			
Date	**Participant**	**Type of participant**	**Level of mediation**
17/09/2011	Subject B	Student	Direct control verbal response
Contents			
M. by watching these videos we can see that they are being used more and more just to say that they are I know the teacher should have guided the students better so that she could teach all the work they do by copying and pasting. Where is the		I'm sure that the technologies used will certainly help them to carry out their research, whether it's at home or at home.	

concern for learning? Where the students are the subjects of their own learning?

Date	Participant	Type of participant	Level of mediation
25/09/2011	Subject B	Student	Direct control verbal response

Contents

J. F. like you, I also think that in the activity shown in video 2 there may be authorship, this will depend on the way the teacher conducts the work after the cut and paste, but I believe that this is where the secret of the good use of technologies in the classroom lies in the way the teacher will use it, this needs to be very well planned.

Date	Participant	Type of participant	Level of mediation
17/09/2011	Subject B	Student	Direct control verbal response

Contents

Watching the situations presented in videos 1 and 2, you realise that none of them will provide students with authorship. In the first situation, the students simply copy and paste into Word, and it appears that this is how they will assemble their work. In the second situation, the teacher has ordered the students to select the information, photos and pages they consider important and paste them into Word. If the work is done in this way, there will be no authorship, but it can be seen that the students are researching in pairs in this situation. If, after the research, she conducts the work in such a way that they read what they have researched, debate the subject and then construct their own texts, then there will be authorship. I don't think it was clearly explained how the newspaper would be produced after the research. For Moran, technologies help us to find information and it is essential to master the search tools, know how to interpret what you choose and adapt it.

to the context. Many people are content with the first results of a research project. They think that if they read it, they already understand it. Research is the first step towards understanding, comparing, choosing, contextualising and applying in some way. That's why simply copying and pasting won't lead to authorship or the construction of knowledge.

Date	Participant	Type of participant	Level of mediation
20/09/2011	Subject B	Student	Direct control verbal response

Contents

Colleague M., it's very clear that, as you said in the first video, the teacher is proposing extremely mechanical work, where the student won't build any knowledge, there will be no authorship or collaboration. I also realised that in the second situation, the students are in pairs and it looks like they're discussing. In this work, the teacher has asked them to select the important information and put it in Word, and how the newspaper will be made after the research, she

Date	Participant	Type of participant	Level of mediation
28/09/2011	Mediator	Mediator	Direct Control Procedural Question
Contents			
That's right, girls. Both positions are correct, but was there collaboration? What do you define as collaboration? Shall we discuss?			

(The top of the first table row reads: "hasn't explained.")

Chart 5 - Map of mediations carried out for subject B in September/2011

Source: author

We noticed that **subject B**'s interactions continue at the level of direct control in the form of verbal responses to questions posed in the "authoring and collaboration with technologies" forum. In accordance with the task proposed by the tutor during this period of the course. The interactions reflect the student's position on the subject of the videos analysed during this period of the course. The mediator, on the other hand, intervenes with direct control in the form of a procedural question. Chart 6, illustrated below, gives an overview of the interactions and mediations that took place in the third period of the course, in which subject B took part.

Map of Mediation			
Participant: Subject B		**Period: 01/10/2011 to 31/10/2011**	
Discussion: Authorship and collaboration with technologies.			
Date	Participant	Type of participant	Level of mediation
01/10/2011	Subject B	Student	Self-regulation
Contents			
After the chat, some concepts became easier to understand. Collaboration is the contribution of individuals who want to achieve something. Co-operation is working together to achieve a common goal. Interaction is the exchange of opinions and knowledge between the individual and the object of knowledge, which can be another individual, a computer, etc. Interactivity is the exchange between individuals mediated by ICT. Authorship is the construction of a work-text, after reading, researching and reflecting on the subject in question. The concepts covered in chat, according to authors. Collaboration according to Barros (1994) involves working collectively to achieve a common goal, this concept is associated with sharing with others. For the author, the concept of co-operation is more complex in that collaboration is included in co-operation. Cooperation is a work of co-realisation. According to Mello (1989) interaction is the basic and initial element responsible for opening a communication channel. Tijiboy (1998) says that through interaction individuals discuss their point of view, get to know and think about different questions and reflect on their own thinking. For Picança, interactivity generates a lot of communication,			

a lot of exchange and a lot of participation between individuals mediated by technologies.			
Date	**Participant**	**Type of participant**	**Level of mediation**
09/10/2011	Mediator	Mediator	Direct directive control
Contents			
Excellent J. One question: is there a difference between co-operation and collaboration? What do you think? Take your synthesis to the wiki material along with the references. Cheers.			

Chart 6 - Map of mediations carried out for subject B in October/2011

Source: author

We can see from the mediation map that **subject B** showed an intemalisation of the concepts covered. This can be seen from the self-regulation type interaction raised in the mapping carried out by the system. It can be seen that the student reflected on the concepts of collaboration and co-operation, presenting theoretical references. The mediator carried out his pedagogical intervention by direct control in the form of a directive, requesting a summary of the material posted by **subject B.**

The following tables show the results of applying the mediation map to **subject C.** Table 7 gives an overview of the interactions that took place during the first period of the course.

Map of Mediation			
Participant: Subject C		**Period: 01/08/2011 to 31/08/2011**	
Discussion: Teaching and learning process: today at school!			
Date	**Participant**	**Type of participant**	**Level of mediation**
17/08/2011	Subject C	Student	Direct control verbal response
Contents			
Nowadays, most teachers don't switch to a teaching-learning process where the student is educated as a whole, taking into account their needs and interests, because this requires: time, which teachers often don't have, such as those who work 60 hours; and a willingness to change, because it's more comfortable to stay as it is, with ready-made texts from the textbook chosen for the year.			
Date	**Participant**	**Type of participant**	**Level of mediation**
20/08/2011	Mediator	Mediator	Direct control model
Contents			
Hi J. That's an important reflection, and already brings up some of the questions I've posted in this forum. Teachers have their share of responsibility for not wanting to retrain, but there are several other factors that make the process difficult. You know more than anyone how difficult it is to be doing a course			

with the heavy workload we have. You are examples to our society.

Source: author

During this period, **subject C** makes contributions to the "teaching and learning process: today at school" forum. We note that in this period the student carries out a majority of direct control type interactions in the form of verbal responses to questions in the forum. The mediator makes direct control interventions in the form of modelling.

Chart 8 shows the results obtained from the application of the mediation map for **subject B in the** interactions carried out in the second period of mapping the subject's data.

Map of Mediation			
Participant: Subject C		**Period: 01/09/2011 to 30/09/2011**	
Discussion: Authorship and collaboration with technologies.			
Date	**Participant**	**Type of participant**	**Level of mediation**
20/09/2011	Subject C	Student	Direct control verbal response
Contents			
In both videos there are signs of collaboration. In the first, the teacher collaborates with the students to find the websites for the research. In the second video, collaboration takes place between the students who form groups. But in both cases there is no collaborative construction because construction is about learning and the simple fact of copying and pasting doesn't generate a learning situation. As for the question of authorship, I can't tell just by looking at the images in the videos because I don't know if the material collected by these students is authored and if it is reliable.			
Date	**Participant**	**Type of participant**	**Level of mediation**
28/09/2011	Mediator	Mediator	Indirect control indirect guidance
Contents			
J. A. V. L. and W. This item brought up another term that hadn't appeared in the discussions: interactivity. What is the difference between interaction and interactivity? What do the authors have to say about it? This week we'll focus on these discussions. Cheers.			

Source: author

The table summarising **Subject B**'s interactions during this period of the course shows that the student continues to make contributions to the forum through direct control interactions in the form of verbal responses. The mediator made indirect control interventions in the form of indirect guidance.The last period of mapping of **subject C**'s interactions is shown in Table 9.

Map of Mediation			
Participant: Subject C		**Period: 01/10/2011 to 31/10/2011**	
Discussion: Authorship and collaboration with technologies.			
Date	**Participant**	**Type of participant**	**Level of mediation**
01/10/2011	Subject C	Student	Self-regulation
Contents			
Interaction - is communication that takes place between two or more individuals. It can be a simple conversation or an exchange of information. Interactivity - is interaction mediated by ICTs.			
Date	**Participant**	**Type of participant**	**Level of mediation**
03/10/2011	Mediator	Mediator	Direct control model
Contents			
Congratulations J. Great synthesis of the concepts. Just to make it even clearer, interaction can be as simple as you described or as complex, in which there is something more than just exchanges, such as the negotiation of different points of view, which is the concept of co-operation that you covered in your post.			
Date	**Participant**	**Type of participant**	**Level of mediation**
02/10/2011	Subject C	Student	Indirect control procedural question
Contents			
L. I went to the pontocom magazine page and was surprised to realise that according to Silva (1998) the concept of interactivity comes from pop art, when there was a fusion of subject and object, and he gave the example of Hélio Oiticica's parangolés. Colleague, these parangolés were banners that the spectator manipulated and wore in performances, at which point he was part of the work, everyone was reading the work. Putting this concept into the technological world makes sense, because each user manipulates in their own way, especially in research, the user makes their own way, without a predefined direction.			
Date	**Participant**	**Type of participant**	**Level of mediation**
03/10/2011	Mediator	Mediator	Direct control model
Contents			
And that's L. and J. Interaction encompasses collaboration and co-operation. Interactivity requires a technological interface. Pop art emerged in Brazil between the 1960s and 1970s, practically at the same time as the advent of technology. J. if I'm wrong let me know, because you're the art education expert.			

Chart 9 - Map of mediations carried out for subject C in October/2011

Source: author

During this period, the student's interactions showed signs of self-regulation. From reflections on interaction in the forum "Authorship and collaboration with technologies". The mediator makes a pedagogical intervention of direct control in the form of a model, using theoretical references to guide the discussion on the topic. **Subject C** follows his contributions, but now an indirect control interaction arises in the form of a procedural question. Finally, we move on to present the results of the application of subject D's mediation map. We begin with Table 10, which illustrates the interactions carried out by the student during the first period of analysis applied to the subject.

Map of Mediation			
Participant: Subject D			**Period: 01/08/2011 to 31/08/2011**
Discussion: Teaching and learning process: today at school!			
Date	Participant	Type of participant	Level of mediation
19/08/2011	Subject D	Student	Direct control verbal response
Contents			
Paulo Freire says. The only person who learns is the one who appropriates what has been learnt, transforming it into what has been learnt, and can therefore reinvent it, the one who is able to apply what has been learnt to concrete existing situations. Various pedagogical trends have emerged over time. Many of these liberal, renewed traditional and technicist trends have not taken on board significant transformations in society. It presupposes the adaptation of individuals to the values and norms in force in class society, through the development of individual culture.			
Date	Participant	Type of participant	Level of mediation
19/08/2011	Subject D	Student	Indirect control targeted question
Contents			
Dear Jessica. For Vieira (2002) it is a process of forming capable and intelligent men. Man must be able to face and solve problems, to seek solutions to resolve situations. The teaching-learning process is a dialectical unity between instruction and education.			
equal characteristics between teaching and learning. According to Kupfer (1997), the teacher must realise that he or she cannot be a controller, relinquishing the position of power that has been conferred on him or her.			
Date	Participant	Type of participant	Level of mediation
20/08/2011	Mediator	Mediator	Direct control order
Contents			

Great W. Bringing these authors to enrich our forum. Put the source of these authors so we can research more on the subject. Cheers J.			
Date	**Participant**	**Type of participant**	**Level of mediation**
20/08/2011	Mediator	Mediator	Direct control order
Contents			
Hum, reflecting on your rich contribution, I thought I'd raise some new questions for our reflection. How do we assess students according to their capacity for effort? Won't this harm them in the future? Aren't we devaluing the student who makes an effort? Cheers J.			
Date	**Participant**	**Type of participant**	**Level of mediation**
24/08/2011	Subject D	Student	Direct control verbal response
Contents			
Teaching is a systematic way of transmitting knowledge used by people to instruct and educate their fellow human beings, usually in places known as schools "Teaching, which is instruction, addresses the intellect and enriches it. Education targets the feelings and brings them under the control of the will. In this way, one can acquire an excellent character of behaviour with little instruction, which already allows one to live happily. On the other hand, a bad character of behaviour can be cultivated without any education, which will be all the worse the more education there is.			
Date	**Participant**	**Type of participant**	**Level of mediation**
24/08/2011	Mediator	Mediator	Direct control direct question
Contents			
Guys. This week we're going to follow the discussion started by Waldir on the differentiation between teaching and learning, as well as the elements of a learning environment? Let's align theory and practice in our discussions, as most of the participants have already done. Cheers J.			

Chart 10 - Map of mediations carried out for subject D in August/2011
Source: author

We noticed that **subject D** begins his contributions to the "teaching and learning process: today at school" forum by resorting to direct control in the form of verbal responses to posts made by other colleagues and the mediator. The mediator makes pedagogical interventions of the direct control type in the form of orders, asking the student to indicate references on the subjects covered in the discussions. **Subject D** continues his interactions through direct control in the form of verbal answers to questions. On the other hand, the mediator resorts to direct control in the form of direct questions to subject **D.**

The next table shows the interactions between **subject D** and the other course participants in the "authoring and collaboration with technologies" forum. Chart 11 is used to describe the development of the discussion and the levels of mediation inferred.

Map of Mediation

Discussion: Authorship and collaboration with technologies.

Date	Participant	Type of participant	Level of mediation
15/09/2011	Subject D	Student	Direct control verbal response

Contents

Is there authorship in the situations presented in videos 1 and 2? If so, what kind of authorship? By whom? And if not, why? Both videos only report on activities ordered by the teacher. These are non-cooperative activities, where the students don't intervene, they just use the Internet to copy information from websites, without being told to read it, analyse it and then transfer it to the Word text editor. If you think about authorship, it doesn't exist, because "the texts and images" are taken directly from the websites without any checking by the students.

Date	Participant	Type of participant	Level of mediation
17/09/2011	Subject D	Student	Self-regulation

Contents

Hello F. We can't think that this is a computer course within the school, because it isn't. Nor can we deny the work being proposed by the teacher at the school. Nor can we deny the work being proposed by the teacher at school. It's valid, even if it's copy and paste. There is indeed an activity being carried out using computers, but it is poorly orientated and needs to be re-studied. Sometimes
We can shred these activities as useless, but we have to be careful not to think we know everything. How many times have we done something like this?

Date	Participant	Type of participant	Level of mediation
17/09/2011	Subject D	Student	Direct self-control Speech directed at the mediator

Contents

It was already in our study material that there are two situations: the teachers who resist and the government that wants to apply these new technologies, just to say it's modernising education. An interesting lesson can be done without the use of state-of-the-art resources as long as it has heart. Teach with affection. Generally, technical resources make teachers more rational and cold towards their students. Anyone who doesn't intend to change will be at a disadvantage, because everything is changing, just look at the videos from the past few weeks.

Date	Participant	Type of participant	Level of mediation
19/09/2011	Subject D	Student	Direct self-control Speech directed at the mediator

Contents

Dear M. Your analysis was very prudent, since it doesn't disregard the activities that were done in the computer lab. The part of the video selected gives an idea, but not to the point of labelling all the work as plagiarism or useless. We can't deny the work being proposed by the teacher at school. It's valid, even if it's copy and paste. There is indeed an activity being done with the use of computers, but it is poorly orientated and needs to be re-studied. Sometimes we can shred these activities as being useless.

Date	Participant	Type of participant	Level of mediation
28/09/2011	Mediator	Mediator	Direct control order
Contents			
Hi W. Your point is very important. The videos are not computer courses and we have to be careful with our posts. The work can be valuable insofar as the information copied can be discussed in the classroom by means of a debate or a production based on the material collected. We can't pass judgement, because everything is valid. What we can do is change our practices, maximising the use of technology. Cheers J.			

Date	Participant	Type of participant	Level of mediation
28/09/2011	Mediator	Mediator	Indirect control indirect guidance
Contents			
M. and W. Both very prudent. We can't label. Let's continue discussing our concepts. The differentiation between interaction and interactivity, as well as co-operation and collaboration. OK? Cheers J.			

Date	Participant	Type of participant	Level of mediation
28/09/2011	Subject D	Student	Direct self-control Speech directed at the mediator
Contents			
Thus, interaction is a mutual, reciprocal or influential action. For Lemos (2000), "interactivity is a specific case of interaction, digital interactivity, understood as a dialogue between man and machine through graphic interfaces in real time". However, for Lévy (1999), "interactivity signals much more of a problem, the need for a new work of observation, conception and evaluation of modes of communication than a simple and univocal characteristic attributable to a specific system", and is therefore not limited to technologies.			

Chart 11 - Map of mediations carried out for subject D in September/2011

Source: author

During this period of the course, **subject D** was very participative, making several contributions to the forum. The first interactions are carried out through direct control in the form of verbal responses to questions raised in the discussion forum. However, the system shows signs of interactions categorised as self-control and self-regulation in the posts made by subject **D.** It can be seen that several interactions of the direct self-control type are made in the form of speech addressed to the mediator. The mediator, in turn, carries out some direct control actions in the form of orders and a sequence of pedagogical interventions of the indirect control type in the form of indirect guidance.Finalising the analysis of **subject D**'s mediation maps, table 12 shows a summary of the interactions carried out by the student in the last period of mapping the levels of mediation of the interactions.

<table>
<tr><td colspan="4" align="center">Map of Mediation</td></tr>
<tr><td colspan="2">Participant: Subject D</td><td colspan="2">Period: 01/10/2011 to 31/10/2011</td></tr>
<tr><td colspan="4">Discussion: Authorship and collaboration with technologies.</td></tr>
<tr><td>Date</td><td>Participant</td><td>Type of participant</td><td>Level of mediation</td></tr>
<tr><td>09/10/2011</td><td>Subject D</td><td>Student</td><td>Self-regulation</td></tr>
<tr><td colspan="4">Contents</td></tr>
<tr><td colspan="4">0 he student's role is important, as they cannot be mere spectators in the face of technological advances. Moran comments that distance education is not just fast food, where the student goes there and is served something ready-made. Distance education is about helping participants to balance personal needs and skills with participation in face-to-face and virtual groups, where we move forward quickly and exchange experiences, doubts and results. In the current teaching model, the teacher occupies the central role.</td></tr>
<tr><td>Date</td><td>Participant</td><td>Type of participant</td><td>Level of mediation</td></tr>
<tr><td>09/11/2011</td><td>Mediator</td><td>Mediator</td><td>Direct control directive question</td></tr>
<tr><td colspan="4">Contents</td></tr>
<tr><td colspan="4">Hi W. Is co-operation greater than collaboration? Discuss here and take your writings to the wiki text, as they are very good. Don't forget to bring in theorists to support these concepts. OK? Cheers J.</td></tr>
<tr><td>Date</td><td>Participant</td><td>Type of participant</td><td>Level of mediation</td></tr>
<tr><td>09/10/2011</td><td>Subject D</td><td>Student</td><td>Indirect control procedural question</td></tr>
<tr><td colspan="4">Contents</td></tr>
<tr><td colspan="4">J. The definitions I've listed have been synthesised from various readings. I didn't realise it was necessary to put an author for each definition?</td></tr>
</table>

Chart 12 - Map of mediations carried out for subject D in October/2011
Source: author

It can be seen from the information in the table that **subject D had** interactions classified in the self-regulation category. The mediator made indirect control interventions, asking procedural questions.

At the end of the presentation of the results obtained from the application of the system in the three periods of mapping the levels of mediation between the participants in the subject of Pedagogical Practice in an ICT-supported environment, we present a complementary report that presents the learning mediation history of the subjects analysed in this research.

The report provides a description of the number of interactions during the three data mapping periods carried out for each of the stages of the course that is the subject of the case study. Values are presented that represent the number of interactions at each level of mediation mapped by the system. The information is presented by grouping the mapping periods, which in this case took place in August, September and October of two thousand and eleven.

At the end of the report there is also a graph summarising the levels of mediation. Each series of the graph represents a level of mediation and the number of interactions mapped at each level. The first learning mediation history presents a scenario of the levels of mediation of the interactions carried out by **subject A in** this research. Figure 7 illustrates this learning mediation history.

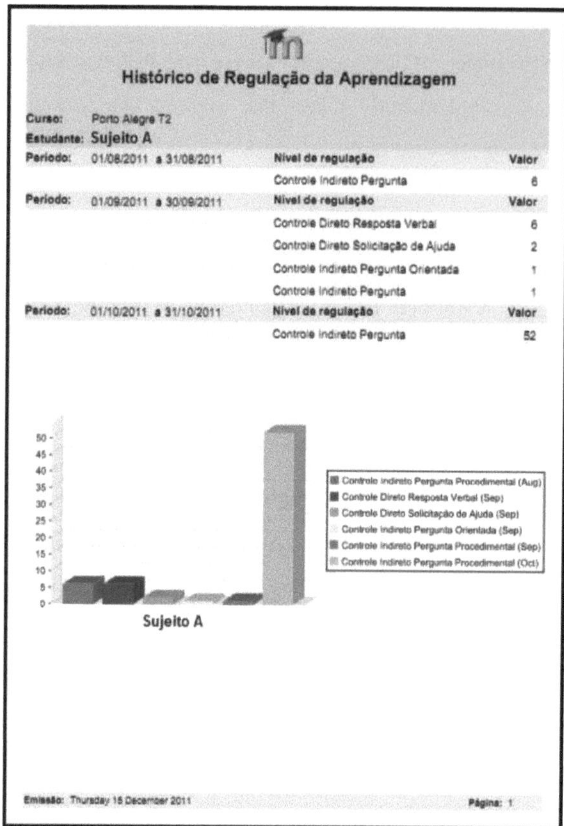

Figura 7   - Subject A's learning mediation history
Source: author

Note that **subject A** resorted to direct control interactions in the form of questions in the first data mapping period. In the second data mapping period, both direct and indirect control were found, with a greater tendency towards direct control interactions in the form of verbal responses. The last mapping period showed a totality of direct control interactions in the form of questions. Figure 8 illustrates **Subject B**'s learning mediation history. We will now describe the information presented in the figure.

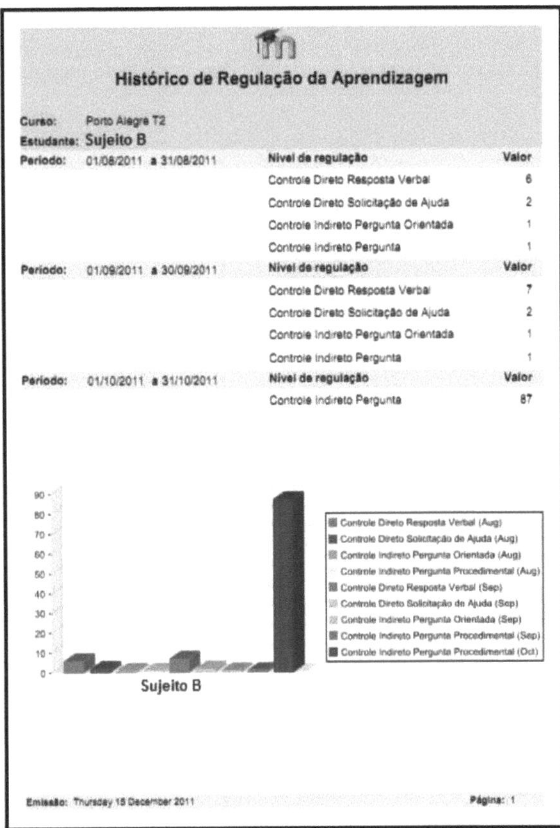

Figure 8 - Subject B's learning mediation history

Source: author

In the first period of mapping the participants' interactions, the system analysed the levels of mediation of the contributions made by **subject B in** the textual interaction tools of the Moodle environment. During this period, the student carried out direct and indirect control interactions, with a higher volume of direct control interactions in the form of verbal responses. In the second period of interaction mapping, direct and indirect control interactions were recorded, as in the first period of mapping. The number of direct control interactions in the form of verbal responses remained higher in this period.

In the last period of mapping the system's data, interactions of the indirect control type were mapped. In this period, all the contributions were in the form of questions. Figure 9 shows an overview of **subject C**'s learning mediation levels based on the set of interactions carried out by the student during the data mapping periods.

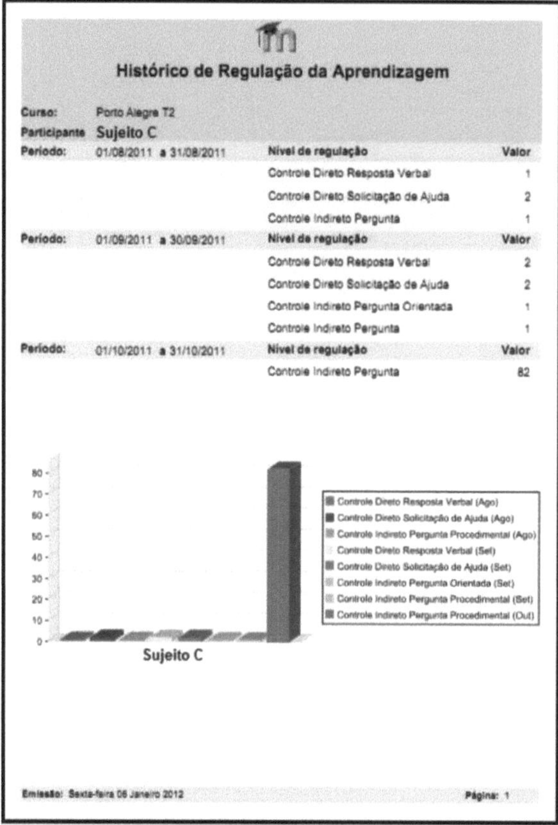

Figura 9   - Subject C's learning mediation history

Source: author

In the first mapping period we can see that direct control type interactions stand out. **Subject C** showed interactions in the form of verbal responses, requests for help and questions. In the second period of mapping interactions, direct and indirect control contributions were noted, with a greater occurrence of direct control interactions. It should be noted that the student used verbal responses and requests for help during this period of the course. In the third period of mapping interactions, all the contributions were at the level of indirect control in the form of questions. Next, we present the learning mediation history of **subject D of** this research. The report is illustrated in Figure 10.

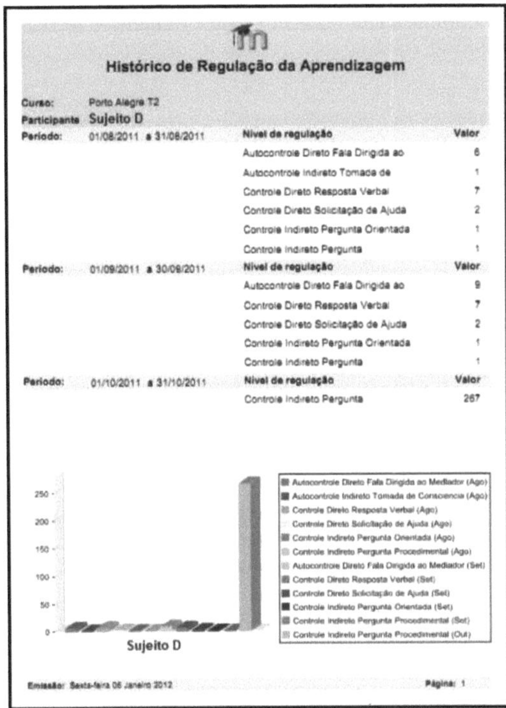

Figure 10 - Subject D's learning mediation history

Source: author

In the first mapping period, we can see from the figure that **subject D** showed control and self-control interactions, in both direct and indirect categories. We can see that there is a greater occurrence of direct control interactions, in the form of verbal responses and requests for help. Followed by an approximate volume of direct self-control interactions in the form of speech directed at the mediator.

In the second period of data mapping, it can be seen that **subject D** carried out a balance of interactions classified in the categories of direct self-control and direct control. The form of direct self-control interaction was speech directed at the mediator. While the most significant form of interaction in the direct control category was a verbal response. We also observed some occurrences of indirect control in the form of guided questions.

The last period of interaction mapping recorded occurrences of the indirect control category in its entirety. The indirect control dimension verified was that of question.

The next section describes the analysis carried out by the researcher himself, who makes observations on the development of interactions during the course from a second point of view.

7.2.3 An analysis by the researcher

In this section, we present an analysis of the results of the application of the system from the point of view of the researcher. From this perspective, we intend to verify the levels of mediation carried out by the participants in the *online* course, based on a check of the students' contributions made in the Moodle environment itself, during the course of Pedagogical Practice in an ICT-supported environment.

The aim of this analysis is to compare the results obtained by the system with the development of the course participants' interactions, as observed by the researcher himself in an analysis of the participants' interactions using the tools of the virtual environment itself. In this case, we analysed the participants' interactions in Moodle's forum, chat and Wiki. These tools were used to carry out the tasks proposed during the three stages of the Pedagogical Practice in an ICT-supported environment course.

In order to complement the analysis of interactions, we also checked the subjects' diaries, since this tool allows students to reflect individually on their own performance. This self-criticism *is* interesting for evaluating the results indicated by the learning mediation survey system.

We therefore used the same methodology to analyse the results obtained from the e-mediation system implemented in the Moodle environment. In this sense, we defined the same subjects as in the system application phase, in order to analyse interactions at this stage of the research.

In addition, we also analysed another aspect relating to user interactions and not just the textual interactions carried out in the tools of the virtual teaching-learning environment. This aspect is related to the accesses to the Moodle environment made by the subjects assessed in the case study. We therefore checked the subjects' accesses and messages in the environment during the course. With this, we sought to identify some characteristics such as time spent in the environment, resources accessed by the course participants, reading material used, contributions made and materials posted by the research subjects. These aspects are not directly related to the mediation of student learning based on textual interactions carried out in the environment, but they do help to monitor subjective aspects that can help the mediator during their pedagogical interventions in the *online environment.*

In order to structure an analysis methodology, we carried out the following steps when observing interactions:

- We observed the interactions that took place based on the tasks proposed at each stage of the Pedagogical Practice in an ICT-supported environment course.

- We observed the interactions of the same subjects mapped in the analysis phase of the system's application.

- We checked forum posts, synchronous meetings held using the chat tool and collaborative writing activities carried out using Moodle's Wiki.

- In parallel with the interactions proposed in the course activities, we analysed the reflections posted by the subjects in their respective diaries. In doing so, we searched for evidence to support the information generated by the learning mediation evidence mapping system.

- We also analysed the data stored in the access *logs to* the Moodle environment during the same period as the course, in an attempt to identify the accesses and messages made by the research subjects.

- We present the results of the analysis for each research subject.

Subject A's accesses during the course, based on Moodle access logs through each subject's individual report, show an overview of the subject's activities in the environment, as shown in Figure 11.

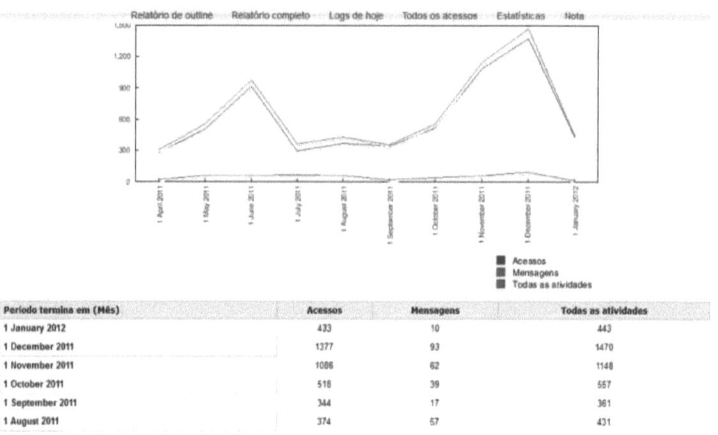

Período termina em (Mês)	Acessos	Mensagens	Todas as atividades
1 January 2012	433	10	443
1 December 2011	1377	93	1470
1 November 2011	1086	62	1148
1 October 2011	518	39	557
1 September 2011	344	17	361
1 August 2011	374	57	431

Figura 11   - Subject A's Moodle access report
Source: https://moodle.cinted.ufrgs.br

During the period in which the data was analysed, from August to October two thousand and eleven, we can see the accesses to Moodle made by **subject A in the** graph shown. We can see that subject **A** made the highest number of accesses in October, when the course was finished, and the lowest number of accesses in September. A greater number of posts were made at the start of the course, in August. However, the greatest number of contributions was made in the first stage of the course. This can be confirmed by the number of posts in the forum, which was highest in August, with ten contributions in this period in the forum: "teaching-learning process: today at school".

Of the contributions made in the forum by subject A in the first stage of the course, the system found that the majority of interactions were classified at the indirect control level. The system identified the form of a question addressed to the mediator as the main level of mediation seen during the period. The mediator's interventions were in the form of directives and modelling.

Subject A's diary allows for a more detailed analysis of the student's own impression of his performance. Figure 12 illustrates the student's diary entries for this period.

Dear diary
I'm having a hard time at first with the subjects
I don't know. I can't find the tasks on the main site, it seems that last semester things were clearer
I think I'm going to delay everything this week, I've only managed one post on the forum, I have to communicate with the tutor

Figura 12 - Subject A's own reflection on the first part of the course
Source: https://moodle.cinted.ufrgs.br

We can clearly see that the student encountered difficulties in this first part of the course. This reinforces the need for greater attention on the part of the tutor in developing her mediating actions. This corroborates the indications raised by the mapping system applied in this research. The system indicated the level of indirect control as the category of student learning mediation based on the mapping of their interactions in the first period of the course. In addition, the interventions carried out by the mediator were direct control in the form of directives and modelling, in an attempt to guide the student through two difficulties.

The second period of the course, which took place in September, was where the student had the fewest activities. However, the number of contributions to the forum was the highest in relation to the three stages of the course. During this period, **subject A** made seven contributions to the forum: "Authorship and collaboration with technologies".

In the second stage of the course, the system indicated a variation in the level of mediation of their learning, with greater evidence of direct control over their contributions in the forum. The main levels of mediation identified by the mapping system in this period were in the form of verbal responses and requests for help. The mediator's interventions were direct control in the form of procedural questions.

We went on to check **Subject A**'s diary entries for the second period of the course. With this, we aim to analyse in more detail the student's own impression of his performance. Figure 13 illustrates the student's diary entries for this period.

Dear Diary...
I was a bit confused by the videos at first, but after participating in the forum I realised that I hadn't understood the idea of the activity correctly.
Well, we see a teacher who comes in with the lesson ready and 'tells the students what to do'. She uses the words "Today we're going", but only the students 'go' to do the research, she takes a vigilant stance, and the interaction about the construction of knowledge is left to the students themselves. I don't see her using a pedagogical method, she just acts as an instructor for the activity, because not even the research sites could be chosen, in other words, everything was formatted.
Once again we can see that technology, in this case, is a new tool that *doesn't* change the methodology, i.e. the use of new equipment to mask an old way of teaching, because there is no effective student participation in a meaningful teaching and learning process.
It's strange how, with all the conditions on offer, certain professionals are unable to evolve their practice.
Bye.
Dear Diary...

I'm at a bit of a loss as to which diary to follow, but here we go.
I was very unwell last week, but I've been trying to keep up with my activities.
I see that every day, every task, I'm evolving in my knowledge, and this is starting to have a positive impact on my teaching practice.
My assessment of the course? Very good!
It's no longer a postgraduate course. It's a space for the production, development and dissemination of knowledge. Congratulations to all of us and, above all, to those responsible for the course.
Bye.
Dear Diary...
I'm worried about the group work. I can't open the tutorial videos on the subject.
To top it off, the tutor said that something was missing from the first post of this second part
What do I do? I think I'm going to cry... Just kidding!
As for the videos, I'm trying them on another computer, but the revision of the first activity of this second part, well, I have to revise it, literally, I thought I had fulfilled the objectives, but now I'm confused.

Figura 13   - Subject A's own reflection in the second part of the course
Source: https://moodle.cinted.ufrgs.br

In this second part of the course, we noticed that the student was concerned about her performance, as she found it difficult to access the course resources in the Moodle environment. In addition, the student was confused about some of the activities carried out during the period. This is evidence of the learning mediation category in the form of requests for help and verbal responses posted by the student in her interactions with the forum tool.

Analysing the *chat* session held with the class during this period of the course, we can see the difficulties presented by the student. We can identify interactions at the level of direct control in the form of requests for help with doubts that arose during the development of the activities.

The mediator's interventions were aimed at guiding **subject A,** through direct control interventions in the form of procedural questions. This can be seen in the evidence provided by the mapping system.

The month of October, when the third and final period for analysing interactions took place, saw the greatest number of activities by **subject A. However, this was the period when the student contributed the least.** However, it was the period in which, proportionally, there were the fewest contributions from the student. During this period, two contributions were posted in the "Authorship and collaboration with technologies" forum.

However, there is evidence of a level of learning mediation in the form of self-regulation. This can be seen in **subject A'**s reflection, illustrated in Figure 14.

My learning during this course was very significant.
During this semester we, the students, have analysed in depth issues relating to the insertion of new technologies in the classroom. Various realities were analysed, and what is available today could be used how and where. The new technologies should be useful, a new possibility for the production and dissemination of knowledge, not merely a different tool for use in a degraded practice without a theoretical basis.
Each learner has his or her own rhythm, so each group of learners also, however small the similarities, has a similar rhythm, not each individual as such, but each group within the reality in which it is inserted, which leads us to believe that there is no point in cutting-edge technology in a place where mobile phones are still a novelty. There's even less point in bringing retrograde production software to a community where social networks are part of students' daily lives. So we can say that even if we work with the old blackboard, chalkboard, notebooks or laptops, it won't do any good if we don't keep abreast of the reality in which our students are inserted, and that doesn't change, it depends on the teacher's commitment. Therefore, the potential, or limitation, of using new technologies in the classroom depends on our knowledge, as mediators of the teaching and learning process, about the content, the technologies and, above all, about our students, our awareness of what the place I'm in is capable of providing and achieving, because if we don't teach for the reality, for the lives of our students, knowledge will play a supporting role in the life history of these human beings, and what we want is the opposite, to mark our passage through the student, a future critical citizen, with the significance of knowledge

Figura 14   - Subject A's own reflection in the third part of the course

Source: https://moodle.cinted.ufrgs.br

We can see signs of self-regulation in **subject A**'s reflections through his entries in the diary tool. There is evidence of the internalisation of the concepts discussed throughout the course, as indicated by the mapping system when it classified the textual interactions of the period in the self-regulation class.

We went on to analyse **Subject B**'s accesses during the three stages of the course. We therefore present below an analysis of the access logs to the Moodle environment through **Subject B**'s report. Figure 15 illustrates the **activities carried out by the student during the period.**

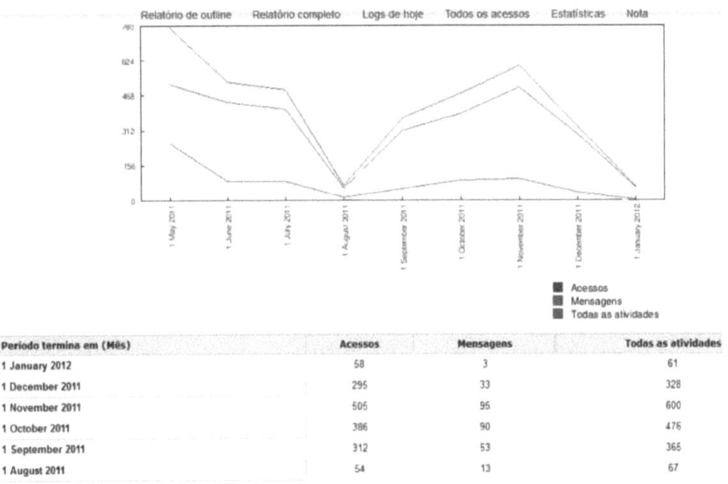

Período termina em (Mês)	Acessos	Mensagens	Todas as atividades
1 January 2012	58	3	61
1 December 2011	295	33	328
1 November 2011	505	95	600
1 October 2011	386	90	476
1 September 2011	312	53	365
1 August 2011	54	13	67

Figura 15   - Subject B's Moodle access report

Source: https://moodle.cinted.ufrgs.br

Subject B's accesses during the interaction mapping period, which took place between August and October two thousand and eleven, can be seen in the graph above. The illustration shows an increasing number of activities carried out by the student during the course. We noticed a smaller number of activities at the beginning of the course, in August. However, the number of activities increased in the last period of data mapping, in October. The period with the highest number of contributions was in October, when twelve messages were posted on the "teaching and learning: today at school" forum. The lowest number of contributions was in October, with only two contributions to the forum.

Let's take a look at **subject B's** own impression of her performance during this period. Figure 16 shows a set of entries made by the student in her diary. An analysis of the posts made by the student in the forum reveals the occurrence of questions in the indirect control category, as indicated by the mapping system.

I believe that my commitment and participation in this first part of the course was very good, I did the reading indicated, I took part in the discussion forum. I think I have achieved the objectives of this first part.
In this first week, I was really struck by the text on the teaching and learning process that we had to read. This text addressed problems that the teaching and learning process is currently facing, it revisited the subject and what interested me most was the paragraph on assessment, which is the regulator of this process.
In my opinion, this second week of studies was very productive. I took part in the discussions in the forums, which I think is very interesting because we interact with our colleagues and exchange opinions. In my opinion, the part that stood out this week was the reflection we had to make based on the video on what teaching and learning is, which enables us to "look" at our practice and rethink it.
In this third week of studies, I really enjoyed watching the videos, but the one I liked the most was methodology and technology, because it made me reflect on what often happens in schools: technology comes along and teachers use it anyway without changing their teaching methods.
In the fourth week of my studies, I consider participation in the forum to be very important and enriching. From my point of view, the exchanges I've had with colleagues have contributed a lot. I try to discuss issues on the forum with my colleagues and I really enjoy it when the exchange takes place. Once I've posted a comment or even commented on a colleague's reply, I look forward to seeing their reaction and also that of others. I believe that this forum is "bearing good fruit".

Figura 16   - Subject B's own reflection on the first part of the course

Source: https://moodle.cinted.ufrgs.br

However, most of the interactions were classified by the system as direct control in the form of verbal responses to the questions posted on the forum. When we check the development of the discussions held on the forum, we can see that the vast majority of posts were made in the verbal response category, as indicated by the system. This confirms the student's level of mediation at this stage of the course in the direct control class. The mediator resorted to direct control pedagogical interventions in the form of modelling.

In the second period of data analysis, referring to the second stage of the course held in September. The student interacted twelve times in the forums "teaching and learning process: today at school" and "authorship and collaboration with technologies". Figure 17 shows **subject B**'s own account in his diary of his performance in this second period of the course.

During these two weeks of study, the exchanges made on the basis of the videos watched were very important for our learning. It's very interesting to comment on the opinions of our colleagues and also to be able to put forward what we think about the subject being discussed.
In this third week of studies, the discussions held in the forum after reading the texts and also after chatting were very enriching for our learning.
For me, these last few weeks working with the wiki have been very important. I didn't know about this tool and after interacting with it, I realised that it's a very interesting resource that teachers can use in their classes. We were able to do a really good job with the contribution of all the colleagues in the group. I thought it was great.
With regard to drawing up the lesson plan, for me it was a very productive activity because we exchanged ideas and managed to plan very interesting lessons, it's very good because we're producing and learning at the same time.

Figure 17 - Subject B's own reflection on the second part of the course

Source: https://moodle.cinted.ufrgs.br

It can be seen from the student's statement that there were many interactions between the participants during this period of the course. The majority of **Subject B**'s interactions were responses to questions posed by colleagues and the mediator. We observed the use of direct control in the form of verbal responses, as indicated by the results of the system's application. Some interactions at the level of direct control in the form of asking for help were identified by the system, which we can see in the student's report, given that the Wiki tool was new to her. The system also identified interactions at the indirect control level in the form of questions directed at the mediator, which we can see by analysing the *chat* with the class.

In the third period of the course, the student made two posts in the "authoring and collaboration with technologies" forum. Figure 18 illustrates **Subject B**'s statement in this part of the course.

In this 3ª part of the course we were offered different discussions in the forum, which I consider very important for our knowledge building, because we have the opportunity to put our opinion regardless of whether it's right or not and on top of this our colleagues put theirs. this provokes important reflections I believe that in this way we learn a lot because we reflect on our thinking, changing it or not.

I thought it was great to build the glossary collaboratively, because this way we're working with tools that we can then use with our students, and as it worked in our experience, we're better prepared to use it in our classroom.

I really enjoyed this course, I believe it provided everyone with significant learning because it was developed using a very diversified methodology.

Figura 18 - Subject B's own reflection in the third part of the course
Source: https://moodle.cinted.ufrgs.br

In this phase, the mapping system identified indirect control interactions in the form of questions directed at the mediator. Most of these interactions took place in the fortnightly *chat* with the class, since only two contributions were made by the student in the forum. Looking at the class *chat,* we can see the presence of guided questions, which led to the results obtained during the mapping of interactions carried out by the system during this period of the course.

We will now analyse the accesses to the Moodle environment made by **subject C.** To do this, we will now report on the activities carried out by the student during the period, as illustrated in Figure 19.

The illustration shows an overview of the activities carried out by **subject C** during the data mapping period, which took place between August and October of two thousand and eleven. We noticed a growing number of activities carried out by the student during the course. Starting with a smaller set of activities in the first month of the course and culminating in a greater number of activities in October, when we finalised the data collection process. The student posted the fewest contributions in the forums analysed during the three periods of the course.

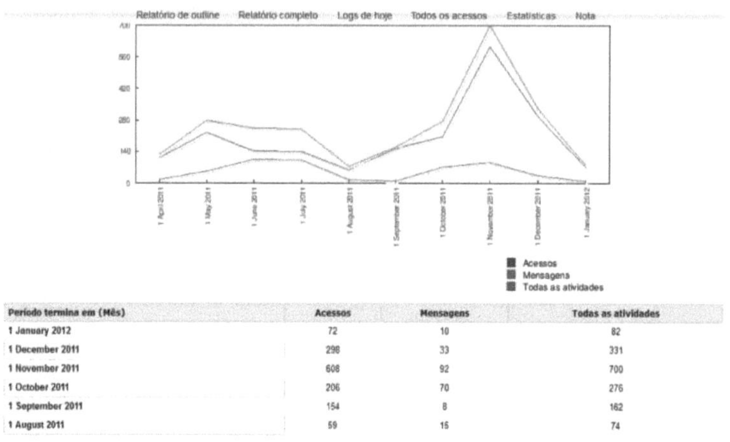

Período termina em (Mês)	Acessos	Mensagens	Todas as atividades
1 January 2012	72	10	82
1 December 2011	298	33	331
1 November 2011	608	92	700
1 October 2011	206	70	276
1 September 2011	154	8	162
1 August 2011	59	15	74

Figura 19 - Subject C's Moodle access report

Source: https://moodle.cinted.ufrgs.br

Looking at **subject C**'s participation in the tasks proposed in the three periods of the course, we noticed greater interaction and contribution from the student in September, in the second part of the course. At this stage, the student made five posts in the "teaching and learning process: today at school" forum. Proportionately, this was the longest period of interaction compared to the others. In the other stages, the student made only two contributions to the forums. Next, we analyse **subject C**'s records of her performance during this period of the course. Figure 20 shows a set of entries made by the student in the diary tool.

I'm late with my reflections, I read the texts, watched the videos in the first week and posted my conclusions on the forum, but I had difficulty with the Gestalt principles and then I got stuck, I couldn't do the exercise for this item, so I took a break.
During this second week, I was able to re-read the design texts and their fundamentals, and put them into practice on the sites I had chosen. Looking at the components of the site as figures, as images, wasn't easy. There are so many factors to observe that at first I was confused, but gradually the lessons were processed in my memory and I was able to really learn. Teaching and learning is very complex. I had no idea that in order to build a simple web page, so many factors had to be taken into account, from colours to space and the layout of texts and images. Any page I built came very much from my notion of aesthetics and not taking into account so many of the criteria I'm learning.

Figure 20 - Subject C's own reflection on the first part of the course

Source: https://moodle.cinted.ufrgs.br

By analysing the forum in the activities proposed for the period, we were able to observe what was identified by the software during the textual interaction mapping stage. The student made only two contributions to the forum proposed as an activity in the second part of the course. The student's interactions were classified as direct control in the form of verbal responses to the questions raised in the forum. The mediator made direct control interventions in the form of modelling, seeking to guide the student in the development of her studies.

We noticed that the student used the diary tool more to reflect on the tasks proposed in the subject. A reading of the diary entries shows that the student has detailed the development of her learning and the difficulties she has encountered. This is in line with the need for help detected by the mapping system.

Figure 21, illustrated below, shows **subject C**'s account of his performance in the second part of the course, where tasks were proposed in the forums on "the teaching and learning process: today at school" and "authorship and collaboration with technologies".

I've been researching how to build a wiki. At first I didn't understand how to do it in the chat, but the next day I reread the chat and then I understood, I made the links and got in touch with the members of my group via email.

The next day I thought about the introduction, and when I accessed the page, my colleague Marcos had already started, so I reread it and contributed some thoughts to complement what he had said

I received a message from my colleague Marcos that the posts in the text were about the concepts of interaction and authorship, but that our tutor Jossi said that the introduction was OK.

I then started researching the concepts that would have to be included in the development and I analysed and posted them. I sent a message to my colleagues saying that they could play with this material, contributing texts or even redoing it, turning this environment into a place for learning and exchange, but so far there has been no contact. My colleague Fátima said in a message that she was building a concept map, and I asked her to send me a copy so that I could also contribute, but she said she hadn't managed to post it, but that she was asking for help.

Figure 21 - Subject C's own reflection on the second part of the course

Source: https://moodle.cinted.ufrgs.br

Of the contributions posted on the forum by the student during this period, there were interactions at the level of direct control in the form of verbal responses to the questions posted on the forums. Analysing the forum posts, we noticed that **subject C had** doubts about the use of the technological resources proposed for the development of the tasks at this stage of the course. The student resorted to direct control in the form of asking for help, which can be observed through the interactions made in the fortnightly *chat* with the class. The student also interacted in the form of procedural questions addressed to the mediator. Mediation was carried out through direct control in the form of orders and indirect control in the form of indirect guidance.

Subject C's participation in the last stage of the course is analysed below. Figure 22 gives us an overview of the student's own statement about her performance during the period.

I wish I'd taken part in the forums more, I think I failed on this point. Teachers have a lot of commitments, especially art teachers, because they're required to take part in many interdisciplinary projects that are developed in the school, the bad thing is that I enjoy it and so do the students. I'm developing a scenario for the end-of-year theatre and the parents' ball, and there have been many others that have unfolded during the year, which I sometimes do outside of my periods together with the students This year we had the MERCOSUR Biennial, which even though it was included in my planning, required visits in the opposite shift
Next year I'll have to plan my time better so that I can devote myself more to the course. My internet has also had problems, I've accumulated a lot of files, I have to make a backup and get into the habit of making copies of all the files, but you only learn this when problems arise.

I've always tried to do my exercises thinking about how I can apply this knowledge I'm acquiring

Figure 22 - Subject C's own reflection on the third part of the course

Source: https://moodle.cinted.ufrgs.br

In this final stage of the course, the student made only two contributions to the "authoring and collaboration with technologies" forum, an activity proposed for the period. These interactions were classified at the indirect control level, in the form of procedural questions directed at the mediator. An analysis of her own statement makes it clear that the student was aware of her lack of activity in the subject. The mediator carried out direct control interactions in the form of modelling.

The next graph, illustrated in Figure 23, shows the activity report for **subject D. We will now** analyse the activities carried out by the student during this period.

Subject D was the most participative of the four subjects observed during the research. The graph shows a greater volume of messages from the student in the three stages of the course.

From the volume of interactions shown in the graph, we see a higher proportion of messages in the first term of the course, held in August. However, the highest number of contributions to the forum was in September, when the student made twenty-four interactions, a period in which the student was proportionally more participative.

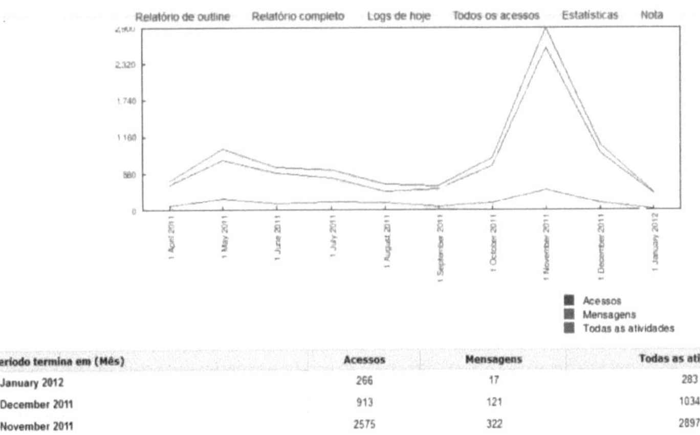

Periodo termina em (Mês)	Acessos	Mensagens	Todas as atividades
1 January 2012	266	17	283
1 December 2011	913	121	1034
1 November 2011	2575	322	2897
1 October 2011	712	118	830
1 September 2011	336	55	391
1 August 2011	297	114	411

Figure 23 - Subject D's Moodle access report

Source: https://moodle.cinted.ufrgs.br

Next, we'll analyse the student's participation and performance, based on his own impressions through his entries in the diary tool during the three periods of the course. Figure 24 illustrates the student's entries in the first period of the course.

I watched the video 'Learning to learn', which shows how we can pass on knowledge through love I answered the proposed questions about teaching and learning and the elements of learning I found it interesting that my material posted on the forum triggered discussions. I made comments on the attitudes of my fellow students 0 another piece of work was related to "Design and Ergonomics", with the painting of images in paint and participation in the forum with her post and the justifications for the use of colours, based on the theory studied. We had to comment on the activity of some colleagues. I didn't find this proposal interesting, since each painting was subjective and each person justified their proposal. It was only if they didn't follow the theoretical ® parameters that I thought they were wrong.

Figura 24   - Subject D's own reflection on the first part of the course

Source: https://moodle.cinted.ufrgs.br

When we analysed the interactions in the forums proposed in the period's activities, we noticed a balance between the contributions made by **subject D.** A set of seven direct control interactions in the form of verbal responses to the questions raised by the course participants. And a set of contributions classified at the level of direct self-control in the form of speech addressed to the mediator. Observing the interactions allowed us to confirm what had been identified by the mapping system.

The student is already at a level of self-control in the first period of the course. Some interactions at the level of direct control were observed, such as: asking for help and orientated questions to the mediator.

However, we also observed some at the level of indirect self-control in the form of awareness.

We noticed few interventions by the mediator in relation to the student. Some direct control interventions in the form of orders were observed. But these orders were related to requests from the mediator to the student about the references for the positions posted on the forum.

The student's statement in the diary allows us to confirm the student's current level of learning mediation. We can see a high degree of autonomy in the development of his learning, characterising self-control.

Next, we analyse **subject D**'s activities, based on his participation in the second stage of the course. Figure 25 illustrates the student's statement about his performance during the period.

(Part 2 - Authorship, interactivity and collaboration: new ways of learning with technologies). Reading and analysing the material: "The computer as a teaching resource", with approaches from Piaget, Vygosky and Paulo Freire on the importance of interaction and the use of computers in education, with opinions from Valente and Papert. I analysed and reflected on the two videos of lessons in the computer lab, observing the previous discussions on the teaching-learning process and teaching resources. The analysis was posted in the diary for the second part. I self-evaluated the first part and posted it in the diary. I think I got ahead of myself with an activity (week 6) on authorship and collaboration with technologies, still related to the videos made at the school in Minas Gerais. I read the material on interfaces: "General principles, icons and texts" and developed activity 5: Analysing screens with icons and different text formatting, posted on the corresponding page as an attachment.©©

Figure 25 - Subject D's own reflection on the second part of the course

Source: https://moodle.cinted.ufrgs.br

The balance between the levels of student mediation is maintained in the second stage of the course. In this period we observed nine contributions at the level of direct self-control in the form of speech directed at the mediator and seven contributions at the level of direct control in the form of verbal responses. By observing the participant's interactions in the proposed forums, we can see what has been verified by the mediation level mapping system. **Subject D** is at the level of self-control in the learning mediation process. We noticed a high degree of autonomy, with significant contributions to the activities developed during the period.

We noticed few interventions by the mediator. In the interventions we observed, we saw direct control in the form of orders, even asking the student for bibliographical references for their positions in the forum. We also noticed some indirect control in the form of indirect guidance.

Observation of the student's diary in the second period of the course confirms **D's** degree of autonomy, with positions and descriptions of concepts based on theoretical foundations, as well as reflections on the development of his learning. The student did not take part in the synchronous meeting organised by the teacher using the *chat* tool, so there is no evidence of mediation in this tool.

To finish analysing the activities carried out by **subject D,** we describe his participation in the third period of the course below. Figure 26 gives us a glimpse of the student's own impression of his performance during this period.

```
            Communities and Personal Learning Environments. My participation in this stage was very exciting,
whether it was reading and contributing to the forum or reading about 'instructional design' and carrying out the corresponding activities.
The feedback from colleagues and the tutor guided my inspiration and re-readings of the world of communities and social networks.

The Internet is becoming a way of life, where users are members of one or more social networks (Orkut, Twitter, Facebook, etc.) Social
networking sites generally work on the basis of user profiles - a collection of facts about what a user likes, dislikes, their
interests, hobbies, education, profession or anything else they want to share. These are web pages that facilitate interaction between
members in various locations.
```

Figure 26 - Subject D's own reflection on the third part of the course

Source: https://moodle.cinted.ufrgs.br

In the last period analysed, we saw fewer student contributions. In this case there were only four contributions, the lowest number compared to the other periods of the course.

The mapping system identified a majority of interactions at the direct control level in the form of procedural questions directed at the mediator. There was also a contribution at the self-regulation level. This confirms what was found in the previous stages of the course, where student performance was classified at the self-control level.

The mediator made few interventions in the interactions carried out by **subject D.** Among these interventions, we noticed direct control in the form of questions.

One point to highlight during the process of analysing the interactions of the subjects in the three periods of mapping the levels of mediation of contributions in the textual tools of the Moodle environment is that there was a delay in the discussions. We noticed this delay because the third period of the course was supposed to discuss the topic of social networks and their implications for education. During this period, the forum "Social networks and education - limitations and potential" was to be used for discussions on the subject. However, we noticed that the course participants were still discussing the topic of authorship and collaboration.

In any case, the delay in the discussions did not jeopardise the survey of the levels of mediation of the students' learning during the period, since many questions and contributions were launched in the forums: "teaching and learning process: today at school" and "authorship and collaboration with technologies".

Another aspect observed during the analyses was the use of the diary tool. Initially, the diary was supposed to be used to reflect on the student's own performance in the subject. However, it was also used for other purposes. Students used the diary to record their positions on tasks proposed in the course. The most appropriate tool for such entries should have been the forum, where questions were posed so that the student could take a position on them.

After the considerations raised in the researcher's analysis, we move on to present the results of the analysis of the interactions of the participants in the course that is the subject of our case study from the perspective of the course tutor.

7.2.4 The subject tutor's view

The purpose of this stage was to compare and triangulate the data collected and analysed by the researcher, directly from the interactions carried out by the course participants in the Moodle environment, with the information generated by the e-mediation system in order to identify and verify the signs of mediation identified. Thus, the purpose of the interview was to gather information on the development of the activities proposed in the course, the participation of the class in general in relation to the interactions carried out in the environment, the general performance of the class and the performance of the subjects analysed in this investigation, the forms of interaction of the participants and the types and pedagogical interventions carried out. We also aimed to identify subjective aspects of the activities carried out in the course, such as: possible delays in developing the activities, difficulties encountered with the environment's tools, time taken to develop the activities, difficulties in understanding the concepts covered and the degree of autonomy and prior knowledge of the subjects analysed.

To do this, we carried out a semi-structured interview, using a questionnaire with some closed questions that served as guidelines for the development of the research. Based on the guiding questions, we unfolded them through open questions that arose during the interview, with the aim of exploring the research more comprehensively. In this way, we sought to identify more elements about the teaching-learning process that took place during the period of application of the system for mapping signs of mediation in the *online* environment.

We will now discuss the approach adopted for the dynamics of the interview. The interview was structured around fifteen questions, organised into four parts:

- **Part I:** consisted of a set of questions about the structure of the subject and the class in which the tutor worked.

- **Part II:** in the second part we devised a series of questions to collect data on students in general.

- **Part III:** the third part of the interview included questions about the development of the pedagogical interventions carried out by the tutor during the course.

- **Part IV:** in the fourth part of the interview we asked the tutor about the performance of the subjects analysed in the research during the course. We asked about each subject's learning profile, individual performance, which subjects performed best and the difficulties they encountered in completing the proposed tasks.

- **Part V:** in the fifth and final part of the interview, the system for mapping signs of mediation developed in this research was presented and the tutor's opinions on the usefulness of the system in helping the teacher to monitor the process of pedagogical mediation carried out in the *online* environment were received.

The first part of the interview included four questions. Below are the questions and comments on the interviewee's answers. The question that began the interview was about the structure of the subject Pedagogical Practices in an ICT-supported environment. The interviewee reported that the course was divided into three stages.

The activities of the first stage consisted of forum discussions, in which students had to post at least two contributions and close each week of the first stage with critical reflections in the diary, especially explaining their impressions of their performance over the period.

In the second stage of the course, the students had to journal their reflections on videos about learning with the use of technology during the first four weeks of the stage. Over the next three weeks, students had to post contributions to the discussion forum on the subject of learning with technology, broadening their level of reflection and theoretical background. At the end of this part of the course, the students had to put together a lesson plan using one of the hundred collaborative writing tools presented.

The third stage of the course began with the topic of social networks and personal learning environments. The tasks at this stage involved the construction of glossaries on the topics learnt and student participation in the forum on social networks and their potential and limitations.

The next question was about the deadlines for the development of the tasks proposed in the structure of the course in all its stages. We asked the interviewee if there had been any delays in any of the activities. If so, the interviewee had to state which task and at which stage the delay occurred. In the interviewee's account, she confirmed that there had been a delay in the development of the task involving collaborative writing using Moodle's Wiki tool, in the second stage of the course. According to the tutor, the students had some difficulty using the tool and took a long time to familiarise themselves with it. In addition, there was a delay in the reflections that had to be made in the diary, as the tutor had to constantly check on the students. Another problem was the delay in posting in the discussion forums. The tutor believes that these delays occurred because the forums are weekly.

When asked about the general participation of the class during the proposed tasks and the level of participation of the students, in particular whether the students were interactive, critical or indifferent during the tasks, the tutor replied that the majority of the class was interactive and critical. The interviewee believes that the fact that the class was participative was due to the constant pressure and encouragement during the tasks. In other words, according to the results indicated by the graphs generated by the e-mediation system, the tutor made direct and indirect control interventions during the development of the tasks carried out by the students. According to the tutor, around seventy per cent of the students were interactive; around twenty per cent were average and ten per cent were indifferent.

In the fourth and final question of the first stage of the course, we asked the interviewee about using the right tools for the proposed tasks. We wanted to know if the students were using the right tools according to the tutor's instructions. For example, did the students use the forum to post contributions to the discussions, or did they use another tool such as a diary? The interviewee confirmed that the students used the right tools for the proposed tasks.

The second part of the interview included eight questions. We asked the tutor to name five of the most participative students in the class. At the time, the interviewee listed seven students because, according to her, listing five is a difficult task, given that all seven were equally participative. The tutor reported that one particular student, not listed among the seven, was very participative. However, the student only did what was necessary. Of the seven listed by the tutor, the four subjects previously selected in the case study are included.

Next, we asked the interviewee which students had performed best in the subject. According to the tutor, the best performers during the period were the same seven listed in the previous question.

So we asked the interviewee if the students' performance is in any way related to their greater number of interactions in the environment. Or if the number of interactions has nothing to do with the process. According to the tutor's response, the students with the most interaction are the best, because they are constantly interacting with their colleagues and the tutor, through tools such as forums, *chat and* e-mail. Based on the interviewee's response, we expanded on the question by asking her which tool was used more to answer questions: *chat* or e-mail? The interviewee said that the participants used Moodle's own messaging tool a lot. They also sent messages by e-mail, as well as using the forums for questions.

Next, we asked the interviewee about the degree of autonomy of the students, asking her to identify the five most autonomous in the development of their learning, with greater independence from the tutor during the course. The tutor reported that the same seven listed in the previous two questions were the most autonomous. The interviewee added that, in her opinion, autonomy, participation and performance should be correlated in distance education for effective student learning. Given the interviewee's response, we followed up the question by asking the tutor if less interactive students, but with a good level of prior knowledge, could not perform very well? Has this happened to any student during the course? The interviewee said that this can happen. However, it is an isolated occurrence. She pointed out that there had even been a student who was not very participative, but who performed well. However, it is worth emphasising that the interviewee said that if the student had participated more, they would certainly have performed better.

In the next question, we asked the tutor to identify what kind of difficulties the students had encountered during the course. We wanted to investigate whether Moodle's tools caused difficulties in the students' performance. The interviewee confirmed that there had been difficulties. These included the use of the Wiki tool, especially when building pages and *links*. The students had problems inserting images into a page in the Wiki tool, they didn't understand how to carry out the task. The interviewee pointed out that the problems were with the tool. They also had difficulties with the forums, not in using the tool itself, but because the forums were very long and had many subdivisions. What's more, the forums were weekly with many topics to discuss, which made the task very tiring for the students. Based on the tutor's response, we asked a follow-up question, asking if analysing the interactions from the forums was a strenuous task for both the tutor and the students? The interviewee replied that it was, because in very long forum structures, with several entries and sub-levels, it makes the process of monitoring interactions difficult. This makes it difficult for the tutor to mediate. In addition, it discourages students from making contributions to discussions, causing them to ignore or make few posts in tasks that involve reflection through the environment's forum tool.

The next question asked what types of doubts occurred in the synchronous sessions held using the *chat* tool. The interviewee reported that the students' biggest doubts and questions were about using the Wiki tool. The students also didn't understand the task that had to be carried out in the collaborative writing tool. Initially, the students thought that the topic to be built collaboratively was free and not a production on the concepts studied and discussed in the forums. The tutor reported that the students had difficulty defining what to do in relation to the task of drawing up a lesson plan.

The last question in this second part of the interview referred to the use of the diary, with the aim of checking whether this tool was only used for students to reflect on their performance in the course, or whether it was also used for other purposes. The tutor replied that the diary was used as a resource for free interaction, in which students could also add other elements about the development of the course. In this case, the students used the diary to express their anxieties and impressions about the subject, as well as about other subjects on the course. With the interviewee's response, we asked whether the identification of subjective facts observed from entries in the diary tool helped to identify the student's profile and mediation needs. The interviewee replied that it did, because many particular problems that may be hindering student performance are exposed in the diary entries.

The third part of the interview included three questions, starting with a question about the pedagogical intervention carried out by the tutor during the period. With this question we aimed to identify the forms of mediation in order to compare them with the results indicated by the mapping system. To do this, we listed the types of interventions made by the mediator, gave examples and asked the interviewee to select which ones were applied to the class. The tutor selected the following forms of intervention: orders, directives, direct questions, conceptual questions, procedural questions, gratifications and passive confirmation. We then asked the interviewee which forms of pedagogical intervention she used the most. The tutor replied that she resorted more to directives, direct questions, conceptual questions and gratifications. The tutor's response confirms what was found by the e-mediation system, i.e. the graphs and reports show that the tutor resorted to control actions during her interventions during the course.

The second question asked about the types of interactions carried out by the students that were identified by the tutor. Again, we presented and exemplified the forms of interaction that could be carried out by the students. Next, we asked the interviewee to tick the forms of interaction adopted by the students. The tutor selected all the forms of interaction presented, namely: verbal response, request for help, procedurally orientated question and conceptual orientated question. As the interview continued, we asked the tutor which of the forms of student interaction were most used by the students. In her reply, the tutor said that all the forms were used a lot by the students. It's worth pointing out that, to complement her answer, the tutor commented that when she answered the question it felt like she was interacting with the students again. At that moment, she visualised the moments of interaction with the students during the course of her tutoring activities in the subject of Pedagogical Practices in an ICT-supported environment. We emphasise that the information generated by the e-mediation system confirms the subject tutor's response, since the exact forms of interaction reported by the interviewee were identified.

The next question was about the students' degree of autonomy. We asked the interviewee if she had noticed any students who stood out for their high degree of autonomy, prior knowledge and superior performance. The tutor answered the question by naming the highest performing students in the class. For the tutor, four students had a superior performance, presenting a high degree of autonomy, present and connected in the activities developed in the subject. Of these four students, two of the subjects analysed in this research are present: subjects B and D. The tutor also mentioned five other students with excellent performance. This group of students includes subjects A and C. This can be confirmed by viewing the information contained in the graphs and reports generated by the system for mapping signs of mediation in VLEAs, presented in the section analysing the results of the application of the e-mediation system.

In the fourth part of the interview, we asked specific questions about the performance of the subjects involved in the case study applied in this thesis. To do this, we used four questions. We began by asking the tutor to report on the learning profile of each subject involved in the case study of this research.

As for subject A, despite her experience in schools, the interviewee reported that the student had a very interesting profile. At the start of the course, the student was not very interactive, as she was very quiet, didn't ask many questions and was somewhat shy. The interviewee noticed that she had a lot of initial difficulty using the environment. This was observed in face-to-face classes, as the student relied heavily on her colleagues for guidance on how to use the environment's tools. However, the tutor reported that she was surprised by the student's progress during the course. The student showed a lot of interest throughout the course, trying to complete her tasks and overcome her limitations. In fact, the interviewee added that another colleague who was very interactive at the start of the course no longer accessed the environment during the course. This gave the false impression that the student would excel throughout the course, but the opposite was true. Whilst subject A, despite the initial difficulty encountered, managed to progress throughout the proposed tasks.

We note that the information generated by e-mediation and the analysis carried out by the researcher are in line with the tutor's report. The information generated by the e-mediation system's graphs, maps and reports identified a degree of evolution in the student's performance. We can see that the number and quality of the student's interactions increased over the three periods analysed. Also, the interventions carried out by the tutor are exactly those reported by her, identifying that the student needed mediation at the control level.

Subject B, on the other hand, is also experienced as a teacher and school principal. The interviewee recounted the fact that the student entered as an alternate in the selection process for the course. The student entered the course with a lot of motivation, as she was looking forward to being classified. In terms of performance, she was a very participative student and didn't present many difficulties. From the start of the course, the student questioned the tutor about her activities, trying to check if anything was wrong and if so, she made no effort to resolve it. The interviewee emphasised that the student has a high level of commitment to her performance, always seeking the best possible performance in her tasks.

It should be emphasised that the tutor's report corroborates the information generated by the e-mediation system. The graphs show that the student was very participative, as confirmed by the tutor. In addition, the student's participation was also noted in the researcher's analysis of interactions on Moodle. The mediation maps generated by e-mediation confirm the quality of the student's interactions and the degree of autonomy pointed out by the tutor in the interview.

Subject C also had some difficulties during the course. The student also has a more restrained profile. However, the student worked very hard on her tasks. The student was very participative, asking questions on issues that presented difficulties. During the course of the course, the student sometimes confused which tool to use and recorded her activities incorrectly. However, she sought guidance from the tutor to resolve her doubts about her mistakes. The interviewee pointed out that the student was often surprised when she carried out her tasks, as she had no previous experience of the Moodle environment, yet she still succeeded. Based on the knowledge acquired in the course, the student began to apply the technologies in her pedagogical activities at school. At the end, the interviewee added that the student brought a lot of experience from her teaching practice to the discussions held on the forum tool.

By analysing the results of the application of the e-mediation system, we can see what was reported by the interviewee. The diversity of interaction levels identified by e-mediation through the mediation graph allows us to visualise the student's initial difficulties, as pointed out by the tutor. The tutor resorted to mediation actions at the control level, trying to help the student develop her tasks. In the analysis carried out by the researcher, we can see the difficulties encountered by the student based on her reflections recorded in the diary tool.

Finally, subject D had an extremely autonomous learning profile. The student was very participative and critical in the positions taken in the forum discussions. He even questioned the positions and issues raised by the subject tutor. The interviewee reported that she likes students with this profile because they are challenging. The tutor reported that the student sometimes appeared to be very incisive in his comments, given his profile. The student sought to make a consistent contribution to the discussions, presenting theoretical bases for the issues raised. They never accepted positions without a theoretical basis in the tasks carried out in the forum. The interviewee praised the student in relation to a criticism of the tutor's performance, since the student complained about the delay or lack of mediation given to the student's positions. The tutor reported that this delay was caused by the student's agility, as he was always ahead of the other colleagues. The tutor emphasised that the student always sought to broaden discussions and demanded this of his colleagues. The interviewee reported that she found the student's participation very positive.

With regard to subject D, he was the only student for whom the e-mediation system identified self-control interactions, which is indicative of the degree of student autonomy reported by the tutor. The interviewee's account confirms what was pointed out by e-mediation through the mediation graphs, history and mediation maps. In the analysis carried out by the researcher, we can see the quality of the reflections made by the student using the diary tool. In analysing the application of the e-mediation system, we can see the levels of the student's interactions through the map of the mediations carried out during the data mapping periods.

The next question was about the performance of each subject analysed in all the stages of the subject. The interviewee pointed out that the students who performed best out of the four analysed were subject D and subject B, in that order. The interviewee reported that the two subjects mentioned were the ones who read the materials used in the course the most, as well as the ones who had the fewest difficulties manipulating the environment's tools. In fact, subject D already had prior knowledge of Moodle, as he had already taken some courses in the area of distance education. Subjects A and C also performed very well in the subject. However, their performance was slightly lower than that of subjects B and D.

The next question was about the difficulties encountered by the subjects analysed in the case study. The interviewee said that the main difficulties were in using the tools, since some of them had no previous experience with them. Some struggled to use the environment. However, these difficulties were only observed in relation to subjects A and C. Subjects B and D had no difficulties using Moodle's tools.

The last part of the interview asked five questions about the e-mediation system. With this, we investigated the tutor's impressions of the system's mediation indicators, in order to compare them with the interviewee's account of the performance of the research subjects and the pedagogical interventions she carried out during the ICT-supported pedagogical practices course.

Before presenting the functionalities of the system for mapping signs of mediation, we asked the interviewee about her opinion on an extension to the virtual teaching-learning environment aimed at supporting the tutor in the pedagogical mediation process. The interviewee said yes, because anything that adds functionality to support the mediation process in the *online* environment is welcome. The tutor also reported that Moodle's access graph is not enough to monitor student participation. In addition, the tutor said that the way the functionality is handled is not good.

We then explained the e-mediation system and presented its graphical interface. We also explained the categories of mediation and how they are used to identify the levels of mediation mapped by the e-mediation system. We explained that the system works as an extension to the Moodle environment and that it is coupled to the environment to provide functionalities to support the pedagogical mediation process. We then began to present the graphs and reports generated by the system.

The first graph presented was the mediation graph. During the presentation of the graph, we detail the information presented by the bar graph. We presented the set of interactions carried out by the course students in the three mapping periods carried out in the case study. We describe the data series generated and the types of interactions identified for each student. We also present the data series relating to the types of interventions carried out by the tutor during the mapping periods. The tutor can visualise the variety of mediations carried out during the course, confirming that the types of mediations shown in the graph reflect the pedagogical intervention actions carried out by her.

The interviewee asked if the graph only shows the types of interactions carried out using the forum tool? The researcher replied no, explaining that the mediation categories presented reflect the interactions carried out through all the tools used throughout the course. In addition, the researcher said that the levels of mediation carried out in the forum tool are shown on the mediation map. The interviewee showed interest and a certain curiosity in visualising the mediation maps generated by e-mediation. The researcher said that he would present the results of the mediation maps during his explanation of the system's functionalities.

However, the tutor commented that the mediation graph generates very interesting information, as it provides a panoramic view of the levels of interaction carried out by the students. The number of interactions can be seen, as well as the relationship between the types of interactions carried out by each student. The tutor also commented that she found it interesting to be able to see the types of mediation she carried out during the course, highlighting the variety of mediation actions in each period analysed. At the end, the interviewee commented that she could have used the mediation graph to identify the level of student participation during the course.

Therefore, the researcher asked the tutor if the mediation chart could help support the mediation process in the environment. In addition, the researcher asked if the tutor could identify the best-performing students using the mediation chart. The tutor answered yes, pointing to the series of mediation levels of the best performing student, confirming the usefulness of the information generated by the graph, in line with the interviewee's own findings.

The researcher then presented the report on the history of mediations to the interviewee. The researcher explained how the information is tabulated and presented to the mediator, demonstrating the survey of mediation levels for each mapping period. The tutor commented that the student's mediation history allows the identification of mediation levels for each student and this is relevant information, as it helps to identify mediation needs in the first few weeks of the course.

The next feature presented to the tutor was the mediation map. The researcher then explained the information presented in the report and the evolution of the levels of mediation carried out. An explanation was given on how to present the information so that the tutor could identify the evolution of the mediations throughout the discussions held in the forums. The interviewee confirmed the levels of mediation indicated by e-mediation on the mediation map, identifying and confirming the levels of her interventions throughout the contributions made by the students.

Having finished explaining the functionalities of the e-mediation system, we returned to the interview questions. To continue, we asked the interviewee if the survey of mediation levels based on the categories presented could help the tutor to monitor the pedagogical mediation process in the *online* environment. The interviewee replied that yes, the mediation categories certainly make it possible to identify the student's level of learning regulation, helping the mediator during their intervention actions in the environment.

It's worth pointing out that in the course of presenting the functionalities of the e-mediation system, we presented the graphs, maps and reports generated over the three periods of data mapping. The interviewee, even though she didn't know the e-mediation system or even the mediation categories for identifying the types of participant interactions, was able to identify the level of evolution of student interactions in the environment quite naturally. In addition, the tutor emphasised that the actions indicated by the e-mediation really do indicate the types of mediation carried out by her. In addition, we emphasise that the students identified by the tutor as the best performers in the subject coincide with the subjects identified by the e-mediation system.

The next question was related to the usefulness of a system for mapping signs of mediation in the virtual teaching-learning environment. We asked the interviewee about the usefulness of the e-mediation system. Whether the system produces useful information that can help the tutor monitor the levels of mediation carried out during an *online* course. The interviewee said yes, because a system like e-mediation could help a lot in observing the evolution of the mediations carried out during the course. The interviewee commented that if she had used the system during the course, it would have greatly reduced her work when identifying the need for pedagogical interventions. It's worth noting that the tutor commented that using the system to map signs of mediation can help with both *feedback on* student performance and gathering elements for student assessment. The tutor pointed out that using the system makes it possible to visualise the students' learning process within the environment.

Next, we asked the tutor about the graphical interface of the e-mediation system. We asked about the information generated by the graphs and reports. Is it appropriate and in a presentation format that makes it possible to identify mediation needs between the tutor and the students? In this respect, the tutor replied that she really liked the presentation of the e-mediation. The interviewee pointed out that Moodle has some reports on the interactions of participants in a course, but these reports produce quantitative information. What sets the **e-mediation system apart is that it produces qualitative information about the interactions that take place in the environment.**

The last question put to the interviewee was about possible improvements to the system. We wanted to get some suggestions from the tutor, since she works directly in the mediation process in virtual teaching-learning environments and could contribute to improving the functionalities presented by the e-mediation system. The tutor suggested some improvements in the visual aspect, with regard to the presentation format of the information generated by the graphs and reports.

The interviewee suggested including subtitles to explain the mediation categories presented, as they can often confuse the tutor. Especially for beginner tutors who don't yet have experience with the terminology used. This can lead to confusion when identifying mediation categories. The interviewee mentioned that we could use short explanatory texts with examples of interventions in each mediation category, making it easier for tutors to identify them. The interviewee also suggested improving some of the information in the headings of some reports. At the end of the interview, the researcher thanked the tutor for her participation and her kindness in collaborating with the investigation, emphasising the importance of her report in validating the results obtained in the development of the case study applied in this research project.

7.3 Discussion of results

We therefore set about triangulating the information generated by the system for each subject analysed in the case study with the findings made by the researcher and the subject tutor. We began by evaluating the results of the mapping of indications of mediation for **subject A.**

The results obtained by the system indicate that subject A was very interactive throughout the three periods of the course. She only slacked off in the final stage, where she had few interactions compared to the other periods. The mapping system identified that the student made a lot of use of procedurally orientated questions to the tutor. In addition, there were many verbal response interactions. In the graphs and reports generated by the e-mediation system, we can see a degree of evolution in the student's interactions. This identifies that the student required control intervention actions on the part of the course mediator.

The researcher's analysis shows that the data provided by the mapping system reflects the levels of interaction observed. The student performed well in her activities, with some doubts arising from the use of the collaborative writing tool, resorting to interactions in the form of requests for help. We found this through the synchronous sessions in the *chat* tool and by reading the student's diary. Self-regulation interactions were noted at times, confirming the student's excellent performance. However, she encountered some difficulties in using the environment's resources during the course of her tasks. This led to the mapping of a higher incidence of indirect control.

The interview with the tutor of the course is in line with what was found by the system and the researcher's analysis. During the report on her performance, the tutor said that the student had had a difficult start, due to her inexperience with using the virtual environment. In addition, the student has a restrained profile, with some shyness, making it difficult for her to interact with the tutor at the start of the course. However, she tried to resolve her doubts and succeeded in her tasks.

By presenting the graphs and mediation reports, the tutor identified her intervention actions in relation to the development of the student's learning in the environment. Her actions of direct control over the tasks carried out by the student stand out.

The conclusion we reached when analysing subject A is that she is at an indirect level of control, with a certain amount of autonomy in carrying out her activities. However, the student needs help from the tutor due to some insecurity in handling the virtual environment's resources.

The results obtained from mapping **subject B**'s interactions indicate that the student was very participative during the course. The information generated by the graphs and reports shows that the student made a variety of interactions in the environment. All interactions were classified in the direct or indirect control categories, with indirect control interactions standing out. The information generated shows that there were verbal responses, some requests for help and many indirect control interactions in the form of questions. The diversity of mediation levels mapped indicates the student's profile. The student was very interactive in the environment.

The researcher's analysis found that subject B was indeed very interactive. When we checked the student's contributions in the forum, wiki and *chat* tools, we could see the quality of her interactions. She always tried to contribute to the discussions in the forum, as well as reflecting on her performance in the diary tool. We noticed that the student used very few interactions to ask for help or questions, where most of her contributions were in the form of verbal responses to questions raised in the discussion forums. Therefore, the mapping system revealed a greater occurrence of indirect control based on the interactions made by the student in the environment.

The tutor's statement during the interview confirms the information gathered by the system. According to the tutor, the student had a high degree of autonomy when carrying out her tasks. The student was very self-critical about her performance, always striving for the best possible performance. The student didn't really have any difficulties with the tools in the environment and was able to complete the proposed tasks with ease. The interventions carried out by the tutor were direct and indirect, with more indirect control actions.

The conclusion we reached when analysing subject B is that the student is at an indirect level of control. However, with a greater degree of autonomy in the development of her activities, close to self-control.

With regard to **subject C,** the results obtained by the e-mediation system showed that the student had the fewest interactions compared to the other subjects analysed. In the first period of the course, the student made a lot of use of direct control interactions in the form of requests for help and verbal responses. However, over the course of the tasks carried out in the course, the student's level of interaction evolved to one of indirect control.

During the researcher's analysis of Subject C's interactions, we were able to identify some of the student's initial difficulties in handling the tools in the Moodle environment. Through her reflections in the diary and also in her *chat* logs, she constantly asked for help. This was observed in the first stage of the course. However, we noticed an evolution in the student's learning over the course of the course, with valuable contributions to questions raised in the discussion forums. As a result, the system classified the student at the indirect control level.

In the interview with the course tutor, we realised the results indicated by the e-mediation system. The tutor said that the student had a lot of initial difficulty with the tools in the *online* environment. This was probably due to her lack of prior knowledge of Moodle. The student was very confused about the tasks and tools she had to use at the start of the course. However, she gained experience and managed to overcome the difficulties she encountered while carrying out her tasks in the environment. In her pedagogical interventions, the tutor highlighted the occurrence of control actions, as the student needed more attention to develop her learning in the environment. The tutor reported that, despite the initial difficulties, the student was successful and stood out at the end of the course, bringing experiences from her professional practice.

We therefore conclude that subject C is at an indirect level of control. The student still needs the mediator's control actions when carrying out her tasks in the environment.

Subject D was the most interactive of the four subjects analysed in this investigation. The system indicated that the student was very participative from the start of the course. The interactions mapped by the system identified a high level of autonomy on the part of the student. In the set of interactions carried out by the student, we noted the occurrence of many contributions in the category of indirect control and self-control.

Analysing the interactions carried out by subject D revealed the information indicated by the system. The student was very participative and critical, always seeking to contribute and enrich the discussions on issues raised in the forum. In his contributions, he reflected on the topics covered in the course through consistent positions, based on theoretical references that he made a point of mentioning. Analysing the diary entries confirms the student's autonomy in developing their learning in the environment. We noticed few interactions involving requests for help, the vast majority of which were verbal responses and speech directed at the mediator.

The tutor's statement confirms the information provided by the system. According to the tutor, subject D was the best performing student in the class. He was always looking to collaborate in forum discussions, seeking to improve the quality and depth of the topics covered during the course. The tutor mentioned that the student showed a high degree of autonomy. He also had previous knowledge of the environment and a lot of professional experience.

Thus, based on the results of the system and the analyses carried out, we conclude that subject D is at a level of self-control. Given that the tutor herself reported that few pedagogical interventions were carried out during the course, as the student showed excellent performance.

CHAPTER 8

FINAL CONSIDERATIONS

In order to begin reporting on the conclusions of the work carried out in this research, it should be noted that it involved two lines of enquiry. Firstly, we addressed questions about the use and combination of computer technologies to design a system for mapping signs of student learning mediation in *online* educational environments. Secondly, the application of the designed system in the process of detecting signs of learning mediation based on interactions carried out by participants in a course in an *online* education environment, with the application of a case study.

The first line of research prompted reflection on the design of a system for mapping participants' interactions in a virtual teaching-learning environment. The design of the system was essential because it was the basis for the case study proposed in the methodology. A conceptual and domain model of the application was drawn up based on an analysis of requirements, which identified what is expected of an application for surveying evidence of mediation in a virtual teaching-learning environment. Architectural aspects involved in the system design phase were also taken into account. Another relevant aspect to be taken into account is that the studies were conducted by analysing student interactions during the course of the ICT-supported pedagogical practices course within the postgraduate course in Media in Education. This course was supported by the Moodle virtual environment. Therefore, the data analysed came from this environment, which required the development of an application for Moodle itself in the form of an extension to the software.

As a result, it was necessary to carry out an in-depth study of the standards for developing extensions for the Moodle environment, which were followed during the process of developing the interaction mapping system so that it could be coupled with the virtual teaching-learning environment.

This stage required several decision-making moments about which technologies to adopt when designing the application. More research emerged at this point, as the text mining technology adopted by the mapping system was coded using Java technology, the same technology used to design the system developed in this thesis. However, the technology used to build Moodle is different. It was therefore necessary to find a way of integrating the two technologies, thus guaranteeing interoperability between Moodle and the mapping system. Having overcome the problem of integrating technologies, we moved on to building the application, starting with the specification documents for the interaction mapping system and applying it to the case study.

The second line of enquiry involved studying and reflecting on an epistemology that would guide the process of identifying signs of pedagogical mediation in the Moodle environment. To this end, the socio-historical theory was adopted to monitor the process of mediation between the participants in the *online* course. This required an in-depth study of the concepts involved in this theory, as well as how socio-historical researchers identify how mediation between subjects occurs. In this case, it was necessary to extend the mediation process to assess how it occurs in virtual teaching-learning environments. The categories of mediation form the core of this line of research, through which the studies were conducted to monitor the process of mediation in the virtual environment.

Both investigative axes converged in the methodological approach applied in this research, where theoretical subsidies were combined with computer technologies to find answers to the guiding questions that outlined the research problem involved in this research work:

I.   How can the mediation process be perceived and managed in the different interaction spaces of the virtual teaching-learning environment?

II.   How can text mining be used to map interactions and identify signs of mediation in virtual teaching-learning environments?

III.   How can information from mapping be made available to help teachers with the task of pedagogical mediation in a virtual teaching-learning environment?

The answer to the first question raised is to understand the mediation process based on socio-historical theory, using an approach that allows us to map out how the mediation process takes place in virtual teaching-learning environments. In this way, we can see how the mediation process can take place in the interaction spaces of virtual environments. To do this, we used the concepts of mediation categories proposed by Dias et al. (1993) and Passerino (2005). These works present relevant contributions to mediation according to socio-historical theory, applied in empirical experiments that resulted in the identification of mediation actions and indications of learning mediation. Therefore, these theoretical references were used to extend the understanding of the mediation process to *online* educational environments.

With regard to how the mediation process can be managed in the different spaces of interaction in the *online* environment, *the* system developed in this thesis was used. The system provides a set of mechanisms and a logical structure that allows the mapping of mediation actions and levels of mediation of student learning in the environment. To do this, the system prepares data from interactions in Moodle so that the text mining mechanism can be applied. In this way, the participants' interactions are classified and organised so that the system can extract information on the levels of learning mediation during a course. The mediator's interventions are also classified, providing an overview of the types of interventions carried out throughout the course. With the information generated by the system, the mediator can check the level of mediation of student learning based on the classification of the interactions carried out by the students in the most diverse textual interaction tools made available by the virtual teaching-learning environment.

The answer to the second guiding question of the research lies in the use of text mining technology provided by the Rapidminer API, more precisely in the combination of the Bayesian inference algorithm in a process of classifying the interactions of course participants into mediation categories. By applying the text mining process, the aim was to identify the levels of mediation between students and mediators.

The third question, regarding how information can be made available as a result of the mapping of interactions to help the teacher in their task of pedagogical intervention in the virtual environment, is answered using the reports and graphs generated by the system. The system generates a general graph on the levels of mediation of the interactions of each participant in the course over periods defined by the mediator. It also generates reports mapping the mediations carried out between a particular student and the other participants in the course, giving the mediator a view of the evolution of the mediations that have taken place. And a report on the mediation history based on the interactions carried out by the student over all the mapping periods defined by the mediator.

Based on the answers to the guiding questions, we arrive at an answer to the research problem: **how can text mining technology and socio-historical epistemology provide elements to support pedagogical mediation in virtual teaching-learning environments?**

The information generated by the e-mediation system proved to be efficient for the mediator to obtain a qualitative overview of how the process of mediation and development of student learning takes place over a period of time. These findings are confirmed by the triangulation of the analyses carried out in this research, when the different collection sources were compared.

The validation of the system for mapping student learning mediation was carried out by applying the software in three periods during the Pedagogical Practices course on the basis of a case study. Comparing the results obtained by the software with an analysis by the researcher and the tutor of the course allows us to prove the usefulness of the system proposed in this research in the process of surveying evidence of mediation in virtual teaching-learning environments.

Therefore, with what has been said in these final considerations, it is believed that the objectives have been achieved with the conclusion of this investigative work. Through the use of a system for mapping signs of mediation in *online* education environments, with the adoption of text mining technology, it is possible to gather elements to support the process of pedagogical mediation in virtual teaching-learning environments.

Some limitations identified during the course of the research will be presented below. We also present some prospects for future research arising from the unfolding of the research presented in this thesis.

8.1 Research limitations and developments

When analysing the results from the researcher's point of view, it was noted that the tasks proposed in the course were often carried out using other textual interaction tools in the Moodle environment, different from the tool initially proposed by the teacher. This was observed, for example, with the use of the diary to record contributions on topics previously posted on the forum tool, which was also confirmed by the course tutor.

However, unlike the researcher's impression, the use of the diary was deliberate, since in some tasks the student was asked to reflect on themes developed throughout the course by recording them in the diary tool. It was noted that the use of the diary was not simply to develop impressions about the student's own performance. According to the interviewee, the diary tool was used as a free space to record various aspects observed by the student during the course.

Therefore, the diary was used for a variety of records, including issues related to other subjects and the course itself. As a result, we realised that a possible future project would be to extend the mapping of interactions to data from other interaction tools in the virtual teaching-learning environment. This would guarantee the mapping of signs of mediation throughout the textual interaction space of the environment.

In addition, we noticed a problem arising from the logical structure of the Moodle virtual teaching-learning environment. This problem is related to the impossibility of mapping the mediations between student and mediator based on the interactions carried out by the *chat* tool, since the structure of the Moodle database does not provide for a relationship between the interactions carried out by the participants in a synchronous section. All interactions are stored sequentially, but without any relationship between them. In this way, it is not possible to identify to which *chat* participant a particular message was directed by the mediator. The only way to verify this is by visually analysing the development of the interactions. This makes it difficult to map the evolution of the mediation process between a particular subject and their mediator. This makes it impossible to draw up a mediation map based on the interactions carried out with this tool. For this reason, the mediation map generated by the system only shows the mediations resulting from mapping the interactions carried out in the forum, since the logical structure of the Moodle database keeps references between the textual records made by the participants in the asynchronous discussions.

We therefore suggest that a study be carried out into remodelling the Moodle database so that relationships between *chat* interactions can be stored. Such a relationship could improve the process of gathering evidence, making it possible to visualise the direct relationship between the mediations carried out by the tutor in the *chat* sessions. This would help in the process of pedagogical mediation in the *online* environment, since many questions and discussions about topics in a course are dealt with through the *chat* tool.

Another aspect observed during this research was the system's limitations in terms of mining the content of documents entered into the environment. Both in student documents, resulting from tasks or contributions made by the students. As well as documents deposited by mediators, such as texts and other support materials for developing the subject's content. These documents are stored in various formats, such as doc, pdf, etc. Therefore, as future work, it is suggested that the system's functionalities be extended to map the content of documents inserted into the environment, thus expanding the space for mapping signs of mediation.

The system for mapping signs of mediation has proved its usefulness in identifying the process of pedagogical mediation in the virtual teaching-learning environment. This was proven by the results obtained from the application of the case study involved in this research and by the validation of the system carried out through the triangulation of the analyses presented. However, an increase in the system's functionalities could be realised in future work. Such an increase would be the development of an alert mechanism based on the mapping of mediation signs carried out by the system.

Such an alert system could identify the level of mediation a particular student is at at a given point in the course. The levels would indicate whether the student needs control or self-control interventions, or whether they are already at the self-regulation level. A traffic light could be used as an interface metaphor to identify students' levels of mediation. A red alert would indicate a need for control, a yellow alert would indicate a student already at a level of self-control, while a green alert would indicate a student with a high degree of autonomy at a level of self-regulation. An experiment along these lines was carried out by Severo et. al. (2011b).

Another outcome of this research would be the design of an extension to the mapping system in the form of a multi-agent systems architecture, which would be responsible for the task of supporting the teacher's pedagogical mediation activities in virtual teaching-learning environments. The pedagogical mediation architecture would present an interface for integration with virtual teaching-learning environments so that the results of interaction analysis can be categorised and presented to the teacher. An initial study on an intelligent agent architecture to support the pedagogical mediation process in *online* environments was presented in the work by Severo et. al. (201 la).

Another prospect for future work would be the development of an ontology aimed at describing the domain of pedagogical mediation in virtual teaching-learning environments, based on mediation categories. This would define a specific vocabulary for the mediation process, avoiding ambiguous interpretations when identifying mediation categories in *online* environments. An ontology could represent an accurate description of the knowledge associated with the mediation process, avoiding the problems of natural language semantics. The ontology of the domain of mediation categories could form a common language for pedagogical agents in a support system for the mediation process in *online* environments. Severo (2009a) describes initial research into the development of an ontology for the domain of pedagogical mediation, where a taxonomy of concepts involved in the process was drawn up.

Finally, another aspect to be explored on the basis of the results presented by this research is the evaluation of teachers and tutors. The evidence of mediation provided by the mapping system allows us to monitor the pedagogical intervention actions carried out by those responsible for the course. The system identifies the types and diversity of interventions carried out during the course. This enables the level of engagement with the activities proposed in the course programme by tutors and teachers to be verified. A model assessment system could be devised based on the information generated by the system.

REFERENCES

ADAMS, E. A proposed causal model of vocational teacher stress. Journal of Vocational Education and Training, vol. 53, pp. 223-246, 2001.

ALMEIDA, M. E. B. Educação a distância na internet: abordagens e contribuições dos ambientes digitais de aprendizagem. Educação e Pesquisa, vol. 29, n. 2, pp. 327-340, 2003.

ANDRADE, A. F.; VICARI, R. M. Building a distance learning environment inspired by Vygotsky's socio-interactionist concept. In: SILVA, M. (Org). Online education: theories, practices, legislation and corporate training. São Paulo: Loyola, pp. 257-274, 2003.

ARAÚJO, C. F.; MARQUESI, S. C. Activities in virtual learning environments: quality parameters. In: LITTO, M. F. and FORMIGA, M. (Org). Distance education: the state of the art. São Paulo: Pearson, 2009.

AZEVEDO, B. F. T; BEHAR, P. A.; REATEGUI, E. B. Analysis of discussion forum messages using text mining software. In: XXII Simpósio Brasileiro de Informática na Educação, Aracajú, v. 1, p. 20-29, 2011.

BAKHTIN, M. Marxism and philosophy of language. 1 Iª ed. São Paulo: Hucitec, 2004.

BAQUERO, R. Vygotsky and school learning. Porto Alegre: ArtMed, 1998.

BASSANI, P. S.; BEHAR, P. A. Analysing interactions in virtual learning environments: a possibility for evaluating distance learning. Revista Novas Tecnologias na Educação, v. 4, n° 1, 2006.

BEKKERMAN, R.; EL-YANIV, R.; TISHBY, N.; WINTER, Y. Distributional Word Clusters vs. Words for Text Categorisation. Journal Of Machine Learning Research, v. 3, 2003.

BÖKEMEIER, J.; KOERBER, J. What is the PHP/JavaBridge. Available on the Internet: http://php-java-bridge.sourceforge.net/pjb/ [accessed 15 Aug 2010].

BRUSILOVSKY, P.; PEYLO, C. Adaptive and intelligent web-based educational Systems. International Journal of Artificial Intelligence in Education, 2003.

CASAMAYOR, A.; AMANDI, A.; CAMPO, M. Intelligent assistance for teachers in collaborative e-learning environments. Journal of Computers & Education, vol. 53, pp. 1147-1154, 2009.

COLE, M. Cultural psychology: a once and future discipline? In: J. J. BERMAN (Ed.), Nebraska Symposium on motivation: cross-cultural perspectives, v. 37, pp. 279-335, 1990.

COLE, R. M. Clustering with genetic algorithms. M. Sc., Department of Computer Science, University of Western Australia, 1998.

COUTINHO, L. *Online* learning through course structures. In: LITTO, M. F. and FORMIGA, M. (Org). Distance education: the state of the art. São Paulo: Pearson, 2009.

DANIELS, Harry. Vygotsky and Pedagogy. Loyola. São Paulo. 2003.

DIAZ, R.; NEAL, C.; AMAYA-WILLIAMS, M. Orígenes sociales de la autorregulación. In Moll, L.C (comp). Vygotksy y la educación: connotaciones y aplicaciones de la psicologia sociohistórica en la educación. Buenos Aires: Aique, 1993.

ECO, U. Como se faz uma tese. 23ª ed. São Paulo: Perspectiva, 2010.

EVERS, W. J. G.; TOMIC, W.; BROUWERS, A. A. Burnout among teachers: Students' and teachers' perceptions compared. School Psychology International, vol. 25, pp. 131-148, 2004.

FARIA, E. T. Interatividade e Mediação Pedagógica em Educação a Distância. Thesis (Doctorate in Education) - Pontifical Catholic University of Rio Grande do Sul: Porto Alegre, 2003.

FAVERO, R. V. M.; TAROUCO, L. M. R. Virtual communities: building knowledge through interaction. Revista Novas Tecnologias na Educação, v. 6, n° 1,2008.

FAYYAD, U.; GRINSTEIN, G. G.; WIERSE, A. Information Visualisation in Data Mining Knowledge Discovery. Morgan Kaufmann Publishers, 2002.

FELDMAN, R.; SANGER, J. The Text Mining Handbook: Advanced Approaches in Analysing Unstructured Data. Cambridge University Press, 2007.

FELDMAN, R.; FRESKO, M.; KINAR, Y.; LINDELL, Y.; LIPHSTAT, O.; RAJMAN, M.; SCHLER, Y.; ZAMIR, O. Text mining at the term level. In: Proceedings of the Second European Symposium on Principies of Data Mining and Knowledge Discovery, 1998.

FELDMAN, R.; DAGAN, I. Knowledge discovery in textual databases (KDT). In:

Proceedings of the First International Conference on Knowledge Discovery and Data Mining, 1995.

FERREIRA, T. B., OTSUKA, J. L, ROCHA, H. V. Interface to Assist Formative Assessment in the TelEduc Environment, XIV Simpósio Brasileiro de Informática na Educação - SBIE 2003, Rio de Janeiro, UFRJ, p. 160-169, 2003.

FIALHO, F. A. P.; TORRES, P. L. Distance education: past, present and future. In: LITTO, M. F. and FORMIGA, M. (Org). Distance education: the state of the art. São Paulo: Pearson, 2009.

FRANCO, S. R. K.; COSTA, L. A. C. Virtual learning environments and their constructivist possibilities. Revista Novas Tecnologias na Educação, v. 3, n° 1, 2005.

FRANCO, M. A.; CORDEIRO, L. M.; CASTILLO, R. A. F. O ambiente virtual de aprendizagem e sua incorporação na Unicamp. Educação e Pesquisa, v. 29, n. 2, p. 341-353, 2003.

FRIEDMAN, N.; GEIGER, D.; GOLDSZMIDT, M. Bayesian network classifiers. Machine Leaming, Springer-Verlag: Berlin, v.29, pp. 131-163, 1997.

GALLIMORE, R.; THARP, R. Educational thought in society: teaching, schooling and written discourse. In: MOLL, Luis C. Vygotsky e a educação: implicações pedagógicas da psicologia sócio-histórica. Porto Alegre: Artes Médicas, 1996.

GIL, A. C. Métodos e Técnicas de Pesquisa Social. 5th Edition, São Paulo: Atlas, 1999.

GLUZ, J. C.; PASSERINO, L. M.; VICARI, R. M. A formal model for mediation processes in VLEs. In: XIX SBIE - simpósio brasileiro de informática na educação, Fortaleza, Ceará, 2008.

GRINSTEIN, G. G.; WARD, M. O. Introduction to data visualisation. In: FAYYAD, U.; GRINSTEIN, G. G.; WIERSE, A. (Eds). Information Visualisation in Data Mining and Knowledge Discovery, New York: Morgan Kaufman, pp. 21-45, 2002.

GUO, G., JAIN, A.K., MA W., AND ZHANG, H. Leaming Similarity Measure for Natural Image Retrieval with Relevance Feedback. IEEE Transactions On Neural Networks, V. 13, n. 4, PP. 811-820, 2002.

GUTIERREZ, F.; PRIETO, D. Pedagogical Mediation: Alternative Distance Education. Campinas: Papirus Publishing House, 1991.

GUTHRIE, L.; PUSTEJOVSKY, J.; WILKS, Y.; SLATOR, B. The role of lexicons in natural language processing. Communications of the ACM, v. 39, pp. 63-72, 1996.

HA, S.; BAE, S.; PARK, S. Web mining for distance education. In: IEEE intemational conference on management of innovation and technology, pp. 715-719, 2000.

HAKANEN, J. J.; BAKKER, A. B.; SCHAUFELI, W. B. Bumout and work engagement among teachers. Journal of School Psychology, vol. 46, pp. 496-

513, 2006.

HALL, M.; FRANK, E.; HOLMES, G.; PFAHRINGER, B.; REUTEMANN, P.; WITTEN, I. H. The WEKA Data Mining Software: An Update. SIGKDD Explorations, Volume 11, Issue 1, 2009.

JUNG, C. F. Metodologia para Pesquisa & Desenvolvimento: aplicada a novas tecnologias, produtos e processos. Rio de Janeiro: Axcel Books, 2004.

KENSKI, V. M. Educação e Tecnologias: novo ritmo da informação. São Paulo: Papirus, Iª Edition, 2007.

KORFHAGE, R. Information Retrieval and Storage. New York: Wiley, 1997.

KOCH, S. H. S, MACIEL, M. C. P., PASSERINO, L. M. The mediation in distance leaming: possibilities of mapping the signs. Proceedings of 9* WCCE 2009, Bento Gonçalves, Brazil, pp. 94, 2009.

KOWALSKI, G. Information Retrieval Systems: Theory and Implementation. Kluwer Academic Publishers, 1997.

KOZULIN, A. Psychological Tools: A Sociocultural Approach to Education. Cambridge: Harvard University Press, 1998.

KRAAIJ, W.; POHLMANN, R. Viewing stemming as recall enhancement. In Proceedings of the 19* ACM Conference on Research and Development in Information Retrieval (SIGIR'96), pp. 40-^48,1996.

KYRIACOU, C. Teacher Stress; directions for future research. Educational Review, vol. 53, pp. 27-35, 2001.

LEI, X.; PAHL, C.; DONNELLAN, D. An Evaluation Technique for Content Interaction in Web-based Teaching and Leaming Environments. In: Proceedings of 3rd IEEE International Conference on Advanced Leaming Technologies, 2003.

LEONTIEV, A. N. O desenvolvimento do Psiquismo. São Paulo: Centauro, 2005.

LEVY, Pierre. The Technologies of Intelligence - The Future of Thought in the Age of Information Technology. São Paulo, 1996.

LI, Y.; ZHANG, C.; ZHANG, S. Cooperative strategy for Web data mining and cleaning. Applied Artificial Intelligence, 2003.

LITTO, F. M. The current international scenario of distance learning. In: LITTO, M. F. and FORMIGA, M. (Org). Distance education: the state of the art. São Paulo: Pearson, 2009.

LITWIN, E. (Org.). Tecnologias educativas en tiempos de internet. Buenos Aires: Amorrortu editores, 2005.

LITWIN, E. (Org.) Distance Education: Themes for a New Educational Agenda. ArtMed, Porto Alegre, 2001.

LUAN, J. Data mining, knowledge management in higher education, potential applications. In: Workshop associate of institutional research international conference, Toronto, pp. 1-18, 2002.

LURIA, A. R. Cognitive development: its cultural and social foundations. São

Paulo: Icon Publishing House, 1994.

MA, Y.; LIU, B.; WONG, C.; YU, P.; LEE, S. Targeting the right students using data mining. In: KDD '00, Proceedings of the sixth ACM SIGKDD intemational conference on Knowledge discovery and data mining, pp. 457-464, 2000.

MACHADO, S. F. Pedagogical Mediation in Virtual Learning Environments. Maringá, PR: 2009. Master's dissertation - State University of Maringá, Postgraduate Programme in Education.

MARTINS, G. A.; LINTZ, A. Guia para elaboração de monografias e trabalhos de conclusão de curso. São Paulo, 2000.

MESSA, W. C. Utilisation of Virtual Learning Environments - VLEs: the search for meaningful learning. Revista Brasileira de Aprendizagem Aberta e a Distância, vol. 9, pp. 01-49, 2010.

MIERSWA, L; WURST, M.; KLINKENBERG, R.; SCHOLZ, M.; EULER, T. YALE: Rapid Prototyping for Complex Data Mining Tasks. In: Proceedings of the 12th ACM SIGKDD International Conference on Knowledge Discovery and Data Mining (KDD-06), 2006.

MOLL, L. C. Vygotsky y la educación: connotaciones y aplicaciones de la psicologia sociohistórica en la educación. Buenos Aires: Aique, 1993.

MOORE, M. The Theory of Transactional Distance. In: MOORE M. G. (Ed.). The Handbook of Distance Education. Mahwah, N.J. Lawrence Erlbaum Associates, pp. 89-108, 2007.

MOORE, M.; KEARSLEY, G. Distance Education: an integrated vision. São Paulo: Cengage Leaming, 2007.

MORAN, J. M. Contribuições para uma Pedagogia da Educação Online. São Paulo: Edições Loyola, 2003.

MORAN, J. M. The new spaces in which teachers work with technologies. In: ROMANOWSKI et. al. (Org.). Local knowledge and universal knowledge: diversity, media and technologies in education. Curitiba: Champagnat, p. 245-254, 2004.

MOTY, B.; FELDMAN, R. Text Mining and Information Extraction. In: MAIMON, O. and ROKACH, L. (Eds.). The Data Mining and Knowledge Discovery Handbook. Springer, 2005.

MySQL. MySQL Reference Manual. Available on the Internet: http://dev.mysql.com [Accessed 08 Mar 2010].

NUNES, I. B. The history of distance learning in the world. In: LITTO, M. F. and FORMIGA, M. (Org). Distance education: the state of the art. São Paulo: Pearson, 2009.

OKADA, A. L. P. A challenge for distance learning: how can collaboration and co-operation emerge in virtual learning environments? In: SILVA, M. (Org). Online education: theories, practices, legislation and corporate training. São Paulo: Loyola, pp. 257-274, 2003.

OLIVEIRA, S. L. Tratado de Metodologia Científica: projetos de pesquisa, TGI,

TCC, monografias, dissertações e teses. 2ª Ed. São Paulo: Pioneira, 2000.

PASSERINO, L. M.; KOCH, S. H. S.; MACIEL, M. C. P.; MARTINS, M. C. C. Mediation through evidence in the linguistic context in virtual learning environments. In: XIX Simpósio Brasileiro de Informática na Educação (SBIE), Fortaleza, CE, 2008.

PASSERINO, L. M. ; GLUZ, J. C. ; VICARI, R. M. MEDIATEC - Technological Mediation in Virtual Spaces to Support Online Teachers. RENOTE. Revista Novas Tecnologias na Educação, v. 5, p. 22, 2007a.

PASSERINO, L. M.; GLUZ, J. C.; VICARI, R. M. A proposal for technological mediation in virtual learning spaces. In: XVIII Brazilian Symposium on Information Technology in Education, 2007, São Paulo. XVIII Brazilian Symposium on Information Technology in Education. Porto Alegre : SBC - Sociedade Brasileira de Computação, v. 1. p. 36-47, 2007b.

PASSERINO, L. M. People with autism in digital learning environments: a study of the processes of social interaction and mediation. Thesis (Doctorate in Informatics in Education) - Federal University of Rio Grande do Sul: Porto Alegre, 2005.

PETERS, Otto. Didactics of distance learning. São Leopoldo: Editora Unisinos, 2001.

QUERTE, T. C. M.; TAROUCO, L. M. R. Support environment for distance education: mediation for co-operative learning. Revista Novas Tecnologias na Educação, v. 1, n° 1, 2003.

RAJMAN, M.; BESANCON, R. Text mining - knowledge extraction from unstructured textual data. In: Proceedings of the 6th Conference of International Federation of Classification Societies, 1998.

RAMINELLI, A.; GLUZ, J.C.; PASSERINO, L.M. Bayesian Model for Classifying Mediation Categories in Distance Learning Tools. São Leopoldo: UNISINOS, 2009. MEDIATEC project technical report, Federal University of Rio Grande do Sul and University of Vale do Rio dos Sinos, 2009.

RAPID-I. RapidMiner Developer Tutorial, 2009. Available on the Internet: rapid-i.com [accessed 08 Mar 2010].

RIJSBERGEN, C. J. V. Information Retrieval. London: Butterworths, 1979.

RODRIGUES, C. A. C. Configurations of pedagogical approaches to distance education. Brazilian Journal of Open and Distance Learning, vol. 10, pp. 71-82, 2011.

ROMANI, L. A. Intermap: a tool for visualising interaction in distance education environments on the web. Dissertation (Master's in Computing), IC/Unicamp, Campinas: 2000.

ROMERO, C.; VENTURA, S. Educational data mining: A survey from 1995 to 2005. Expert Systems with Applications, v. 33, pp. 135-146, 2007.

SALMON, G. E-moderating: the key to teaching and learning online. London: Ed. Kogan Page, 2000.

SALTON, G.; McGILL, M. J. Introduction to Modem Information Retrieval. McGraw-Hill, 1983.

SCHIAFFINO, S.; GARCIA, P.; AMANDI, A. eTeacher: providing personalised assistance to e-leaming students. Journal of Computers & Education, vol. 51, pp. 17444754, 2008.

SEVERO, C. E. P.; PASSERINO, L. M.; GLUZ. J. C.; RAMINELLI, A. Pedagogical mediation in virtual teaching-learning environments through educational data mining agents. Informática na Educação: Teoria e Prática, v. 14, n. 2, 2011a.

SEVERO, C. E. P.; PASSERINO, L. M.; LIMA, J. V. A pedagogical mediation support tool for the Moodle environment. In: Primer MoodleMoot Uruguay, 2011, Montevideo. EI primer MoodleMoot Uruguay, 2011b. v. 1. p. 1-5.

SEVERO, C. E. P. ; PASSERINO, L. M. ; KOCH, S. K. S. ; MACIEL, M. ; GLUZ, J. C. . An ontology for mediation categories according to an epistemological approach based on social interaction. RENOTE. Revista Novas Tecnologias na Educação, v. 7, p. 56, 2009a.

SEVERO, C. E. P. ; PASSERINO, L. M.; LIMA, J. V.. E-mediation: a proposal of a miner tool for virtual leaming environment. In: IFIP World Conference on Computers in Education, 2009, Bento Gonçalves. Technology and Education for a Better World. Porto Alegre : WCCE, 2009b. p. 233-236.

SILVA, A. C.; SILVA, C. M. T. Evaluation of virtual learning environments. In: SILVA, A. C. (Org). Learning in virtual environments and distance education. Porto Alegre: Editora Mediação, p. 73-88, 2009.

SILVA, M. F. Chats and e-forums in virtual distance learning: links between pedagogical mediation and hypertextuality. Dissertation (Master's in Linguistics) - Federal University of Ceará: Fortaleza, 2008.

SMOLKA, A. L. B. O (im)próprio e o (im)pertinente na apropriação das práticas sociais. Cadernos CEDES, Campinas: Cedes, n. 50, p. 26-40, 2000.

SMOLKA, A. L. B.; GOES, M. C.; PINO, A. The constitution of the subject: a recurring question. In: J. WERTSCH, P. DEL RIO & A. ALVAREZ. Socio-cultural studies of the mind. Porto Alegre: Artmed, 1998.

SUMATHI, S.; SIVANANDAM, S. N. Introduction to Data Mining and its Applications, Berlin Heidelberg: Springer-Verlag, pp. 499-627, 2006.

TALA VERA, L.; GAUDIOSO, E. Mining student data to characterize similar behaviour groups in unstructured collaboration spaces. In: Proceedings of the Workshop on Artificial Intelligence in CSCL. 16th European Conference on Artificial Intelligence, (ECAI2004), Valencia, Spain, pp. 17-23, 2004.

TELES, L. E-learning. In: LITTO, M. F. and FORMIGA, M. (Org). Distance education: the state of the art. São Paulo: Pearson, 2009.

TORRES, P. L. FIALHO, F. A. P. Distance education: past, present and future. In: LITTO, M. F. and FORMIGA, M. (Org). Distance education: the state of the art. São Paulo: Pearson, 2009.

VAZ, M. F. R. International standards for the construction of *online* educational material. In: LITTO, M. F. and FORMIGA, M. (Org). Distance education: the state of the art. São Paulo: Pearson, 2009.

VENTURA, S., ROMERO, C., AND HERVAS, C. Analyzing Rule Evaluation Measures with Educational Datasets: A Framework to Help the Teacher, In: Ist International Conference on Educational Data Mining, pp. 177-181, Montreal, 2008.

VIEIRA, R. S. O papel das tecnologias da informação e comunicação na educação a distância: um estudo sobre a percepção do professor/tutor. Brazilian Journal of Open and Distance Learning, vol. 10, pp. 65-70, 2011.

VYGOTSKY, L. S. The Social Formation of the Mind. São Paulo: Martins Fontes, 2007.

VYGOTSKY, L. S. Pensamento e Linguagem. 2. ed. São Paulo: Martins Fontes, 1998.

VYGOTSKY, L. S.; LURIA, A. R. Studies in the history of behaviour: apes, primitive man and children. Porto Alegre: Artes Médicas, 1996.

WERSTCH, J.; RÍO, P. Del; ALVAREZ, A. Estudos Socioculturais da Mente. Porto Alegre: ArtMed, 1998.

WERTSCH, J. La mente en acción. Buenos Aires: Aique, 1999.

XU, D.; WANG, H. Intelligent agent supported personalisation for virtual leaming environments. Journal of Decision Support Systems, vol. 42, pp. 825-843, 2006.

XU, D.; WANG, H.; WANG, M. A conceptual model of personalised virtual leaming environments. Journal of Expert Systems with Applications, vol. 29, pp. 525-534, 2005.

YANG, Y.; PEDERSEN, J. O. A comparative study on feature selection in text categorisation. In: D. H. FISHER (Ed.). Proceedings of ICML'97, 14th International Conference on Machine Leaming, p. 412, San Francisco: Morgan Kaufinann Publishers, 1997.

YIN, R. K. Case Study Research: Design and Methods. 2 ed. Thousand Oaks: Sage, 1994.

I **want** morebooks!

Buy your books fast and straightforward online - at one of world's fastest growing online book stores! Environmentally sound due to Print-on-Demand technologies.

Buy your books online at
www.morebooks.shop

Kaufen Sie Ihre Bücher schnell und unkompliziert online – auf einer der am schnellsten wachsenden Buchhandelsplattformen weltweit! Dank Print-On-Demand umwelt- und ressourcenschonend produziert.

Bücher schneller online kaufen
www.morebooks.shop

Printed by Books on Demand GmbH, Norderstedt / Germany